# Springer Series on Atomic, Optical, and Plasma Physics

Volume 79





For further volumes:
http://www.springer.com/series/411

The Springer Series on Atomic, Optical, and Plasma Physics covers in a comprehensive manner theory and experiment in the entire field of atoms and molecules and their interaction with electromagnetic radiation. Books in the series provide a rich source of new ideas and techniques with wide applications in fields such as chemistry, materials science, astrophysics, surface science, plasma technology, advanced optics, aeronomy, and engineering. Laser physics is a particular connecting theme that has provided much of the continuing impetus for new developments in the field, such as quantum computation and Bose-Einstein condensation. The purpose of the series is to cover the gap between standard undergraduate textbooks and the research literature with emphasis on the fundamental ideas, methods, techniques, and results in the field.

Volodymyr Girka · Igor Girka · Manfred Thumm

# Surface Flute Waves in Plasmas

## Theory and Applications

Springer

Volodymyr Girka
Igor Girka
Department of Physics and Technology
V.N. Karazin Kharkiv National University
Kharkiv
Ukraine

Manfred Thumm
Institute for Pulsed Power and Microwave Technology
Karlsruhe Institute of Technology
Karlsruhe
Germany

ISSN 1615-5653
ISBN 978-3-319-37516-8 ISBN 978-3-319-02027-3 (eBook)
DOI 10.1007/978-3-319-02027-3
Springer Cham Heidelberg New York Dordrecht London


Softcover reprint of the hardcover 1st edition 2014


Printed on acid-free paper

Springer is part of Springer Science+Business Media (www.springer.com)

# Preface

A comprehensive study of the properties of eigen waves propagating across the axis of symmetry in various cylindrical waveguide-structures filled with magneto-active plasma is performed in the present book, using the example of flute waves. The aim of the authors is to collect in a single book the materials devoted to various characteristics of these transverse waves first of all, their dispersion properties and their impact on various design features of waveguides, like possible application of a dielectric coating on the inner wall of the cylindrical waveguide. Their damping due to various mechanisms, like collisions between plasma particles, resonant damping, transformation into the coupled bulk mode, radiation through a narrow slot in the waveguide wall are also considered here. Since a real experimental device confines a non-uniform plasma, the impacts of an inhomogeneity of the plasma density and of a constant external magnetic field on the frequency spectrum and the spatial field distribution of these waves are studied here as well. Changing of the surface flute wave properties due to the influence of the shape of the waveguide cross-section including non-circularity of both constructive elements, namely the plasma column and the metal waveguide, and the coupling between different flute modes propagating in a current-carrying plasma waveguide are also in the scope of our book. We have even demonstrated how one can investigate the properties of surface waves, which in addition to their azimuthal mode number have also a small axial wave number, using the theory of surface flute waves. It should be mentioned that all problems considered in this book are solved by the method of successive approximations. It is surprising that the theory of surface waves propagating strictly across the axis of symmetry in relative simple models of cylindrical waveguides allows one to solve the problems of transverse wave propagation in squared waveguides, and to develop the theory of azimuthally non-symmetric surface waves, which have a small axial wave number. Therefore, the material presented in this book will be useful for graduate students specializing in the field of plasma physics and for professionals who are interested in problems of specifically confined plasma structures.

Interest in plasma physics, observed over the last 50 years, is maintained primarily due to the prospects of energy generation by controlled thermonuclear fusion and owing to astrophysical studies. However, research for solutions in these complex fields also leads to the active development of various applied investigations in related areas of science such as plasma electronics, physics of gas discharges, collective effects of charge carriers in plasmas of solids, plasmonics, and nano-science. The plasmas, which are investigated in these fields of science in order to find some new prospective applications, have different values of characteristic parameters, and hence their properties are different. But common for any plasma is that all internal processes are accompanied by the propagation of electromagnetic waves.

Another unifying feature of plasma phenomena observed on the Earth is the limited space, which is occupied by a plasma. These two factors are common for all problems, which are discussed in this book, and which are belonging to electrodynamics of restricted magneto-active plasmas. This circumstance distinguishes our book from well-known fundamental monographs on plasma electrodynamics. Another fact, which testifies the importance of our book, is connected with the wide field of the presented existing applications of surface wave propagation and prospective ones, namely fabrication and processing of a large-sized semiconductor plates using gas discharges sustained by azimuthally non-symmetric surface waves, production of nano-crystals with outstanding mechanical and electrical properties, elaboration of new bio-sensors, which allow one to perform diagnostics of animal and plant pathogens, to carry out in situ gene analysis or rapid testing of a malarial strain, water purity analysis, etc.

One of the Co-authors, V. O. Girka is indebted to the grant of DAAD, which allowed him to take part in the final preparation of the text for publication as a book at the Karlsruhe Institute for Technology (KIT). He is also very grateful to Prof. J. Jelonnek and colleagues in the High-Power Microwave Division of the Institute for Pulsed Power and Microwave Technology at KIT for a lot of fruitful discussions and their kind hospitality. Volodymyr and Igor Girka thank Dr. I. Pavlenko for his participation at solving definite problems discussed in Chaps. 3 and 5 of this book.

Kharkiv, Ukraine
Karlsruhe, Germany

Volodymyr Girka
Igor Girka
Manfred Thumm

# Contents

# Chapter 1
# Introduction

Considering the possible content of our book we have paid the main attention to the following features. First, this should be new material, which has not been completely and consistently presented in previously published books. Second, this should be a presentation of theoretical results, which have some practical applications at present time and which would be prospective for future utilizations in new branches of physics and related spheres of knowledge. And at third, this book should be interesting and useful for young students who would like to study plasma physics.

This book is devoted to studying the properties of surface flute waves propagating in magneto-active plasma, whose boundary has a finite curvature radius. The applications of surface waves are known during a long time. At first they were described by Lord Rayleigh. In [1] one can find a classification of different types of surface waves. We propose for the present considerations a specific type of surface waves, whose field on the plasma-dielectric interface, has a surface type spatial distribution only in the region occupied by the plasma, but in the dielectric region it has a bulk type distribution. This is the first peculiarity of the material presented in this book. Because of the leakage of wave energy into the dielectric region, we are considering here metal waveguides with plasma filling.

The second peculiarity of our book is connected with the geometry of the magneto-active waveguides, in which these waves can propagate. Most of monographs and reviews, where surface waves are considered, make their theoretical study for the case of planar geometry, or they apply either cylindrical geometry inconsistently, or they consider only the case of non-magnetized plasma [2–5].

The next distinguishing feature of our book is connected with the fact that we study just flute waves. This means that these electromagnetic perturbations have no axial wave number. From one point of view this peculiarity allows one to separate Maxwell equations (in the limiting case of uniformity of the considered plasma system) into two independent sets, and from the other side most papers in the field of plasma physics are devoted to the opposite case, namely to studying azimuthally symmetrical electromagnetic waves. But in any case we restricted the scope of our book just to properties of transverse non-symmetrical eigen waves.


V. Girka et al., *Surface Flute Waves in Plasmas*, Springer Series on Atomic, Optical, and Plasma Physics 79, DOI: 10.1007/978-3-319-02027-3_1,

Nevertheless, in one of the sections we demonstrate that the developed theory can be also successfully applied for consideration of surface waves with a finite value of their axial wave number.

The fourth distinguishing feature of the presented material is connected with the fact that we consider here only the case of magneto-active plasma. Indicating that we should like to mention a physical incident, which happened with the well-known physicist L. Landau and his scholar, A. Akhiezer (everyone who is interested in plasma physics knows the fundamental monograph devoted to electrodynamics of plasmas, which was written by a team of physicists led by A. Akhiezer [6]). L. Landau worked for a long time in Kharkiv, where he has created the kernel of his huge theoretical school. This story was told to two of the co-authors of the book by old scientific collaborators of the Ukrainian Institute for Physics and Technology (present title is National Scientific Center "Kharkiv Institute for Physics and Technology"). Once A. Akhiezer has shown his work to L. Landau. He read it and asked the question to his scholar: "What is the strange choice of the object of your investigation? Alexander, where you have seen the plasma immersed into a magnetic field?" Maybe it was a joke; we were not present at this dialog. But nevertheless, our book just deals with different cases, where the plasma is influenced by an external magnetic field.

The fifth peculiarity of the book is connected with the applied theoretical method. Using the method of successive approximation we have solved a lot of tasks devoted to the investigation of the influence of plasma density non-uniformity, different types of external magnetic field non-uniformity, spatial inhomogeneity of plasma interfaces, etc. on the frequency spectrum of the extraordinary polarized surface flute waves (X-modes). We have considered mainly these modes because the corresponding ordinary polarized waves (O-modes) in the most cases have a bulk feature. Thus the book will be useful for students, who want to study such a popular analytical method like the method of successive approximations, which is widely used not only in plasma physics but also in other branches of physics.

Plasma filled waveguides are actively studied both theoretically and experimentally for the purposes of plasma physics [7], radio-physics [8], and plasma electronics [9]. Experimental investigations of various plasma filled electronic devices show prospective applications of surface wave propagation, and aim in increasing their efficiency. During recent years plasmas are widely used in various designs of antenna systems. It has been proved experimentally that plasma antenna systems have a lot of advantages compared with conventional metal antenna systems [10]. This is not only valid for military applications, but also for civil communication systems. Moreover, the range of their civil applications becomes more and more wider at present time [8]. Waves being excited in restricted plasma waveguide structures should be eigen waves of such structures. Surface waves can propagate not only in the case of gaseous plasmas but also in the case of solid state plasmas, see e.g. [11, 12]. Moreover, they can propagate in multilayered planar waveguide structures where the layers can be produced from meta-materials with different electro-dynamical properties. The possibility of surface wave propagation along a flat plasma-metal boundary was indicated for the first time in [13] and after that their properties have been actively studied by many other authors. A review devoted to properties of surface waves propagating along planar plasma-metal boundary across an external magnetic field is given in [5]. However, in experimental practice one often deals with a plasma boundary, which is characterized by a curvature radius of finite value and this circumstance strongly affects the properties of surface waves propagating in such plasma-waveguide structures. The finite curvature radius leads to changes of the frequency ranges where these waves can propagate, of the spatial distribution of their fields, of their polarization and so on, see e.g. [14].

The propagation of SWs is also widely used in various plasma technologies, which utilize gas discharge for cleaning and polishing solids, for lasers pumping and plasma production, elaboration of fullerenes etc. This can be explained by the fact that these waves possess a lot of advantages over bulk waves, for instance, they can produce plasmas with more spatially uniform density profile, and the efficiency of their interaction with plasma particles is also higher [15]. Surface localized plasmons (special type of surface waves connected with perturbations of conducting electrons in metals) are actively applied for formation of nano-structures using metal plasmas. Progress in elaboration of the bio-sensors based on the surface plasmon resonance technique indicates new prospective areas of surface wave applications.

At present time the world community is actively searching for new sources of energy. The International Thermonuclear Experimental Reactor (ITER) is being built in Cadarache, France. Surface waves can propagate in such fusion devices as well. Their propagation has a double role: on the one hand their excitation in fusion reactor leads to losses of RF power intended for ion cyclotron resonance heating (ICRH) of the fusion plasma core, impurity generation and other undesirable events, but on the other hand they could be utilized for cleaning the inner surface of the reactor vessel.

Construction of more complicated plasma devices unavoidably leads to the appearance of more complicated plasma phenomena, for example, to coupling between different electro-magnetic waves. Experimentalists often deal with plasmas, which are either sustained by a flow of electric current, or when flowing of a current in the confined plasma is necessary for plasma stability, if the plasma is sustained by another method. For instance, in tokamaks the plasma is confined both by toroidal and poloidal magnetic fields, the last one is mainly produced just by the axial electric current. That is why we also studied the case of surface flute waves propagating in current-carrying plasmas. In ordinary practice the value of an axial electric current is sufficiently small to consider the azimuthal magnetic field produced by it as a small parameter of the studied problem. Thus we have restricted our research here to the case of small values of this azimuthal magnetic field.

Chapter 2 of the book is entitled "Surface flute waves propagating in uniform magneto-active plasma filled waveguides". Here the influence of the various waveguide parameters on surface flute wave properties is examined under the condition of uniform plasma filling; the external constant magnetic field is assumed to be uniform and directed strictly along the axis of the cylindrical waveguide. The dependence of the surface flute waves (SFWs) on co-ordinates and time is assumed in the form $\sim \exp(im\varphi - i\omega t)$, where $\varphi$ is the azimuthal angle, $m$ and $\omega$ are the azimuthal mode number and their angular frequency, respectively. Application of such a simple waveguide model allows one to study properties of SFWs in a modal approximation. The main attention is paid to the properties of X-modes, because the flute O- modes are often of bulk type.

It will be shown that flute X-modes can propagate both along plasma-dielectric and plasma-metal boundaries and if the external magnetic field turns to zero. The problems of SFW damping are also studied here. It is proved as well that application of the SFW theory allows one to describe dispersion properties even of the eigen non-symmetrical surface waves, which have a finite value of the axial wave number. The theory of SFWs developed in this Chapter is applied in the following Chapters for solving corresponding problems by the aid of the method of successive approximations.

Chapter 3 considers coupled surface flute waves, which propagate in current-carrying cylindrical plasma waveguides. The need for studying such waveguides is explained by the fact that in order to produce gaseous plasmas in laboratory devices one can utilize an electric current. This is why we consider there the influence of an axial electric current on the dispersion properties of SFWs propagating in waveguides with non-uniform plasma filling. The value of this electric current is assumed to be sufficiently small so that either the azimuthal component of an external magnetic field is much less than its axial component or the cyclotron frequency, which appears due to the azimuthal magnetic field, is much less that the eigen frequency of the flute modes in the case if there is no axial magnetic field. Four different geometries of the utilized waveguide structure are studied there, namely: plasma-filled cylindrical metal waveguide, which has a dielectric coating on its inner surface, cylindrical metal rod immersed into magnetized plasma, cylindrical metal waveguide completely filled with plasma, and cylindrical coaxial waveguide completely filled with plasma.

The results of studying the propagation of SFWs in current-carrying waveguides prove the existence of coupling between O- and X-modes while O-modes can be either of surface or bulk type. In different plasma-filled waveguides, coupling between these flute modes leads to different phenomena: appearance of non-reciprocal frequency ranges in their frequency spectrum, change of the sign of their group velocity, damping of the X-modes and so on. The power transfer between these flute modes is characterized by the nonlinear dependence on the small parameter of the problem, which is proportional to the electric current.

The forth Chapter of this book is devoted to investigation of dispersion properties of SFWs while they propagate across the axis of symmetry in cylindrical metal waveguides with non-uniform magneto-active plasma filling. The non-uniformity can be connected with a radial dependence of the plasma density, which fills the waveguide, or with a spatial non-uniform distribution of the applied magnetic field. This could be a radial non-uniformity of the external magnetic field, a toroidal magnetic field non-uniformity utilized in metal waveguides, which are completely filled with plasma, or a non-uniform toroidal magnetic field created along a metal ring, and a weakly rippled magnetic field.

There the frequencies of SFWs and the spatial distribution of their field are examined and the damping rates, which are connected both with collisions between the plasma particles and resonant conversion of these modes into bulk modes, are calculated. It is interesting that from the mathematical point of view radial density non-uniformity is similar to radial non-uniformity of the external magnetic field. Thus influences of these non-uniformities on the SFW frequency spectrum are also similar. A small toroidicity of the external magnetic field leads to SFWs propagation in the form of wave packets. Corrections to the wave field of their main harmonic and to their eigen frequency are proportional to the squared small parameter of the system in this case. Ripples of the confining magnetic field in modern thermonuclear fusion reactors are characterized by a small parameter $\varepsilon_m << 1$. Nevertheless these ripples can influence the conversion and absorption of SFWs. In this case SFWs propagate as well as wave packets, which have six field components. The dispersion equation obtained in Chap. 4 contains small summands proportional to $\varepsilon_m^2$, thus the frequency correction is also proportional to the same value.

Chapter 5 is entitled "Surface flute waves propagating in waveguides with non-circular cross-section". The most effective way to excite waveguide structures including those, which are corrugated, can be realized for excitation in the range of their eigen frequencies [16]. It is known that the shape of the cross-section of the plasma-vacuum interface in fusion devices also distinguishes from a circular one [7, 17]. In the case of stellarators this peculiarity of the design is particularly well seen. Using an expansion into Fourier series one can simulate any real shape of the cross-section of any plasma device. Thus we have solved the following tasks: propagation of SFWs in corrugated metal waveguides completely filled with plasma and propagation of SFWs in a metal waveguide with non-circular cross-section, which is partially filled with plasma. The influence of a cross-section non-circularity of the plasma column on the dispersion properties of these waves propagating in magneto-active waveguides in both resonant and non-resonant cases have been investigated as well.

Periodical changes of the curvature of the plasma column surface and/or a metal chamber along the azimuthal angle are shown to be the reason for SFWs propagation in the form of wave packets. The SFW eigen frequency correction, which is caused by the deviation of the waveguide cross-section from circular shape, is found to have a small value of second order over the small parameter of the problem in the non-resonant case. But in the resonant cases connected with degeneration of the frequency spectrum, the value of the frequency correction can be proportional to the small parameter of the problem in the first power. Examination of the spectral content of SFWs propagating along such waveguides has also been carried out here.

The sixth Chapter of our book contains a review of the possible applications of surface waves propagation. We tried to show there both existing plasma devices, which apply surface wave propagation and prospective possible future applications of surface waves studied in the previous Chapters. These applications concern the following areas: plasma electronics, including devices utilized for generation of super high frequency radiation, plasma-antenna systems including designs, which utilize plasmas sustained by surface wave propagation, plasma production, especially plasma sources based on surface waves propagation, surface plasmon polaritons propagating in the super high frequency range, plasmonics devices, including bio-sensors, photovoltaic cells etc. (see, e.g. [18–20]), and plasma nano-science. The cases of surface wave propagation in thermonuclear reactors are also considered here and the possibility to utilize SFWs for cleaning the inner walls of large-scaled fusion reactor vessels supplied with superconducting magnetic coils is discussed.

## References


1. Moisan, M., Shivarova, A., & Trivelpiece, A. W. (1982). Experimental investigations of the propagation of surface waves along a plasma column. *Plasma Physics, 20*(11), 1331–1400.
2. Gradov, O. M., & Stenflo, L. (1983). Linear theory of a cold bounded plasma. *Physics Reports (Rev. Sec. Phys. Lett.), 94*, 111–137.
3. Alexandrov, A. F., Bogdankevich, L. S., & Rukhadze, A. A. (1984). *Principles of plasma electrodynamics*. Berlin: Springer-Verlag.
4. Vukovic, V. (1986). *Surface waves in plasmas and solids*. Singapore: World Scientific.
5. Azarenkov, N. A., & Ostrikov, K. N. (1991). Surface magnetoplasma waves at the interface between a plasma-like medium and a metal in a Voigt geometry. *Physics Reports, 308*, 333–428.
6. Akhiezer, A. I., Akhiezer, I. A., Polovin, R. V., Sitenko, A. G., & Stepanov, K. N. (1975). *Plasma electrodynamics*. Oxford: Pergamon Press.
7. Wesson, J. A., & Lashmore-Davies, C. N. (1997). *Tokamaks*. Oxford: Clarendon Press.
8. Anderson, T. (2011). *Plasma antennas*. Boston: Artech House.
9. Kondratenko, A. M., & Kuklin, V. M. (1988). *Basis of plasma electronics*. Moscow: Energoatomizdat. (in Russian).
10. Jenn, D. C. (2003). *Plasma antennas: Survey of techniques and the current state of the art.* San Diego:SPAWAR PMW 189.
11. Beletski, N. N., Bulgakov, A. A., Khankina, S. I., & Yakovenko, V. M. (1984). *Plasma Instabilities and non-linear phenomena in semi-conductors*. Kyiv: Naukova dumka. (in Russian).
12. Agranvich, V. M., & Mills, D. L. (1982). *Surface polaritons*. Amsterdam: North-Holland.
13. Toda, M. (1964). Propagation of waves in a solid state plasma waveguide in a transverse magnetic field. *Journal of the Physical Society of Japan, 19*(7), 1126–1130.
14. Girka, V. (2006). Theory of transverse surface electro-magnetic waves propagating along a plasma–metal boundary with a finite radius of curvature in a magnetic field. *Plasma Physics Reports, 32*(5), 401–410.
15. Aliev, Yu M, Schluter, H., & Shivarova, A. (2000). *Guided-wave-produced-plasmas*. New-York: Springer.
16. Kuzelev, V. M., & Rukhadze, A. A. (1990). *Electrodynamics of dense electron beams in plasma*. Moscow: Nauka. (in Russian).
17. Volkov, E. D., Suprunenko, V. A., & Shyshkin, O. O. (1983). *Stellarators*. Kyiv: Naukova dumka. (in Russian).
18. Maier, S. A. (2007). *Plasmonics: Fundamentals and applications*. New-York: Springer.
19. Green, R. J., Frazier, R. A., & Shakesheff, K. M. (2000). Surface plasmon resonance analysis of dynamic biological interactions with biomaterials. *Biomaterials, 21*, 1823–1835.
20. Ederra, I., Azcona, L., & Alderman, B. E. J. (2007). A 250 GHz sub-harmonic mixer design using EBG technology. *IEEE Transactions on Antennas and Propagations, 55*(11), 2974–2982.

# Chapter 2
# Surface Flute Waves Propagating in Uniform Magneto-Active Plasma Filled Waveguides

Let us consider in this Chapter the properties of electromagnetic waves propagating across the axial direction in cylindrical metal waveguides filled with magneto-active plasmas. The influence of the waveguide parameters, such as radius of plasma column, value of the external magnetic field, constructional peculiarities etc. on the wave properties is examined here under the condition of uniform plasma filling. This assumption is valid both for the case of solid state plasmas (e.g. plasmas of semi-conductors) and for the case of gaseous plasmas with a thin transient plasma layer, whose thickness is small in comparison with the penetration depth of the studied modes. In the case of waveguides filled with magneto-active plasmas, it is assumed that the external constant magnetic field is uniform and has only an axial component.

The dependence of the electromagnetic fields of the surface flute waves (SFWs) on co-ordinates and time is assumed in the form: $f(r)\exp(im\varphi - i\omega\, t)$, where $\varphi$ is the azimuthal angle, $r$ is the radial co-ordinate, $m$ and $\omega$ are the azimuthal mode number and frequency of the SFW, respectively. In the cold magneto-active plasma approximation, the Maxwell equations can be separated into two independent sets of equations, which describe extraordinary and ordinary polarized waves (or in other words: X- and O-modes, respectively). Such waves connected with the perturbations of charged plasma particles are analogical to gravitational flute modes connected with the motion of plasma density ripples across an external constant magnetic field [1]. Propagation of gravitational flute modes can be unstable under the condition of large radius of the plasma column and/or large plasma density gradient.

Since the radial distribution of O-modes has often a bulk character, and they do not depend on an external axial magnetic field but strongly depend on features of the dielectric layer between plasma column and metal wall of the waveguide, the main attention is given here just to the properties of the X-modes. Thus the next three sections are devoted to studying the dispersion properties of eigen flute X-modes. The properties of the eigen flute O-modes are investigated only in the Sect. 2.4. Features of flute X-modes damping are discussed in the Sect. 2.5. Nevertheless knowledge on the O-modes is necessary for studying the case of coupled flute modes propagating in current-carrying waveguides. This problem is under consideration in the Chap. 3 of this book.

V. Girka et al., *Surface Flute Waves in Plasmas*, Springer Series on Atomic, Optical, and Plasma Physics 79, DOI: 10.1007/978-3-319-02027-3_2,

Here we also show the possibility to apply the theory of surface flute waves to the derivation of a dispersion equation for non-symmetric electromagnetic waves, which in addition to the azimuthal mode number, are characterized by a small axial wave number. It is found that their dispersion equation has two solutions for the case of extraordinary polarized non-symmetric waves, namely fast and slow modes. Here the analogy between these waves and the resistive drift waves described in [1] should be mentioned once more.

## 2.1 Extraordinary Polarized Surface Flute Waves Propagating in Magneto-Active Plasma Waveguides

Let us consider the properties of SFWs, which propagate along the interface of a uniform plasma column with radius $R_1$, which is separated by a dielectric layer from the metal wall of a cylindrical waveguide with radius $R_2 > R_1$. An external constant magnetic field is oriented along the axis of the waveguide $\vec{B}_0 \| \vec{z}$. The scheme of this waveguide is presented in Fig. 2.1. We consider in this section the case of a dense plasma, when the inequality: $\Omega_e^2 > \omega_e^2$ is valid, here $\omega_e$ and $\Omega_e$ are the electron cyclotron and plasma frequencies, respectively.

In the region of the plasma column, the set of Maxwell equations derived for eigen flute X-modes can be reduced to the modified Bessel equation [2] for the SFWs' magnetic field. Consequently, its solution can be represented in the following general form:

$$H_z = A_1 I_m(k_\perp r) + A_2 K_m(k_\perp r), \tag{2.1}$$

where $k_\perp = k\sqrt{(\mu^2 - 1)\varepsilon_1}$, $\mu = \varepsilon_2/\varepsilon_1$, $k = \omega/c$, $A_{1,2}$ are constants of integration, $I_m(z)$ and $K_m(z)$ are the modified Bessel and MacDonald functions, respectively [2], and $\varepsilon_{ik}$ is the dielectric permittivity tensor of a cold magneto-active plasma [3].

Then in the region $R_2 > r > R_1$ the fields of the SFW are described by a superposition of Neumann functions $N_n(z)$ and first kind Bessel functions $J_n(z)$ [2]:

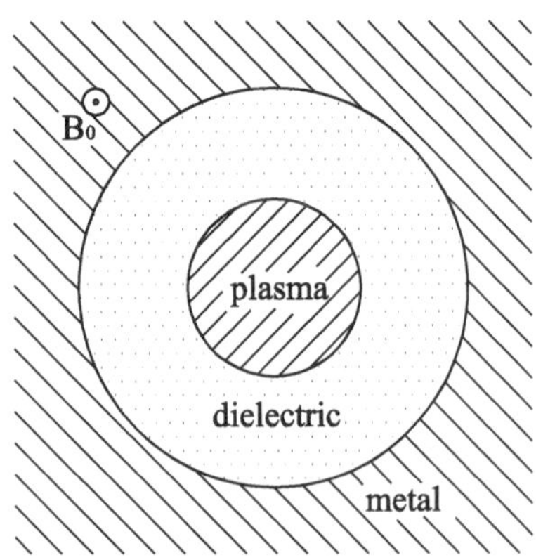


**Fig. 2.1** Scheme of a *metal* waveguide, which is partially filled with plasma

$$H_z = B_1 J_m(kr) + B_2 N_m(kr), \tag{2.2}$$

where $B_{1,2}$ are constants of integration.

To determine the integration constants, one can apply the following boundary conditions

(1) SFW fields should be restricted inside the whole waveguide and, in particular, on the waveguide axis;
(2) Tangential components of the SFW fields should be continuous on the plasma-dielectric interface;
(3) Tangential components of the SFW electric field should be equal to zero on the metal wall of the waveguide.

The X-modes studied here exist in the form of surface type waves if the following inequality is valid: $k_\perp^2 > 0$. The frequency ranges, within this inequality is valid, are determined by the following expressions:

$$\omega_{LH} < \omega < |\omega_e|, \quad |\omega_e| < \omega < \omega_1, \tag{2.3}$$

$$\omega_{UH} < \omega < \omega_2, \tag{2.4}$$

where $\omega_{1,2} = \mp |\omega_e|/2 + \sqrt{\Omega_e^2 + \omega_e^2/4}$ are the cut-off frequencies for bulk modes, and $\omega_{LH}$ and $\omega_{UH}$ are the lower and upper hybrid frequencies, respectively. Let us entitle in this Chapter the frequency ranges (2.3) and (2.4) as the low frequency (LF) and high frequency (HF) ranges, respectively. Applying the boundary conditions indicated above, one can derive the following dispersion equation $D_0(\omega) = 0$ for the eigen flute X-modes:

$$\begin{aligned} D_0 = & \frac{I_m'(k_\perp R_1)}{I_m(k_\perp R_1)}\sqrt{\frac{\varepsilon_1}{\varepsilon_2^2 - \varepsilon_1^2}} + \frac{m\varepsilon_2}{kR_1\left(\varepsilon_2^2 - \varepsilon_1^2\right)} \\ & + \frac{J_m'(kR_1)N_m'(kR_2) - J_m'(kR_2)N_m'(kR_1)}{J_m(kR_1)N_m'(kR_2) - J_m'(kR_2)N_m(kR_1)}, \end{aligned} \tag{2.5}$$

here and in the following text, the prime symbol denotes the derivative of the corresponding cylindrical function over its argument.

Let us consider the most interesting case, when the thickness of the vacuum layer which separates the plasma column from the inner surface of the metal waveguide is sufficiently small that the following inequality is valid: $\Delta = 1 - R_1 R_2^{-1} << 1$. Then the effect of the plasma on the properties of the surface flute X- modes (SFXMs) is the most pronounced. In this case the dispersion equation (2.5) can be rewritten in the more simple form:

$$\frac{I_m'(k_\perp R_1)}{I_m(k_\perp R_1)}\sqrt{\frac{\varepsilon_1}{\varepsilon_2^2 - \varepsilon_1^2}} + \frac{m\varepsilon_2}{kR_1\left(\varepsilon_2^2 - \varepsilon_1^2\right)} \approx \frac{m^2 - k^2R_1^2}{kR_1}\Delta. \tag{2.6}$$

Analyzing equation (2.6) one can see that the azimuthal mode number $m$ cannot be equal to zero if $\Delta \neq 0$; and only in the special case of the plasma-metal boundary when $\Delta = 0$ one can realize oscillations with $m = 0$ [4]. One can obtain simple expressions for the solutions of dispersion equation (2.6) in some limiting cases. They are given in the following formulas.

In the HF range (2.4) the sign of the azimuthal number can be only negative, and for $|m| >> 1$ the eigen flute X-mode frequency can be represented in the following form:

$$\omega \approx \omega_1 \left( 1 + \frac{m^2\delta^2}{2R_1^2} - \Delta m \frac{\omega_e}{\omega_1} \right), \tag{2.7}$$

where $\delta = c/\Omega_e$ is the skin depth and the inequality $1 >> m^2\delta^2 R_1^{-2} >> m\omega_e \Delta\omega_1^{-1}$ is assumed to be valid. In the limiting case of large values of the waveguide radius ($k_\perp R_1 >> |m|$) the flute X-mode frequency can be expressed in another form:

$$\omega \approx \omega_1 + \Omega_e \frac{m^2\delta^2}{2R_1^2} \left( 1 - 2m\Delta \frac{\omega_e}{\omega_1} \right), \tag{2.8}$$

here $R_1^2 > \delta^2 m^2 \Omega_e / |\omega_e|$; $\omega_1 > m\omega_e\Delta$. For the SFXMs, whose frequency belongs to the LF range (2.3), the dispersion equation (2.6) has the following solution:

$$\omega^2 \approx \frac{|m|\,\Delta\Omega_e^2}{1 + |m|\,\Delta} \left\{ 1 + \frac{\Delta R_1^2 \left[1 + 2\Delta\,(1 + |m|)\right]}{2\,(1 + |m|\,\Delta)^2\,\delta^2\,(1 + |m|)} \right\}, \tag{2.9}$$

which is derived in the limiting case $R_1 << \delta$, $\delta\,|m| < 1$. If the inequality $|m| >> 1$ is valid then one can find the following expression for the flute X-modes frequency:

$$\omega \approx \frac{\pm\,|\omega_e|}{2\psi_1} + \sqrt{\frac{\Delta\,|m|\,\Omega_e^2}{\psi_1} + \frac{\omega_e^2}{4\psi_1^2}}, \tag{2.10}$$

here $\psi_1 \approx 1 + \Delta R_1^2 \delta^{-2} |m|^{-1} + 2.7 R_1 \delta^{-1} |m|^{-1}$, the upper sign relates to the case of positive $m$, and the lower sign to the case where $m < 0$. In the limiting case $k_\perp R_1 >> |m|$, the analysis of the dispersion equation (2.6) allows one to find the following solution:

$$\omega \approx \frac{-m\omega_e}{2\psi_2} + \sqrt{\frac{m^2\Omega_e^2\Delta}{\psi_2} + \frac{m^2\omega_e^2}{4\psi_2^2}}, \tag{2.11}$$

here $\psi_2 \approx R_1\delta^{-1} + R_1^2\delta^{-2}\Delta$, $R_1 >> \delta\,|m|$. Thus, the azimuthal mode number $m$ can be both positive and negative in the LF range unlike in the case of HF flute

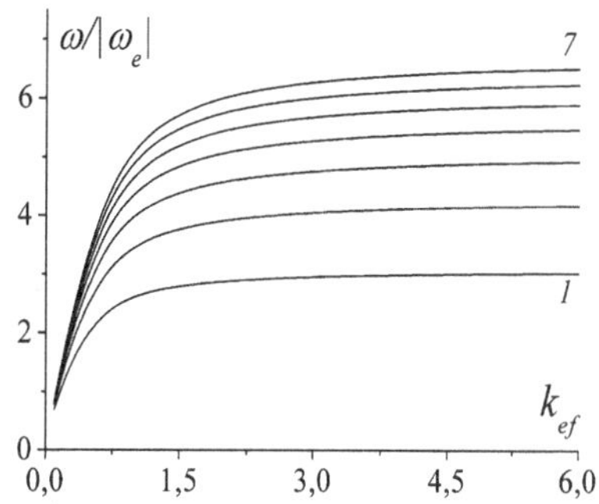


**Fig. 2.2** Frequency of LF SFXMs propagating in a metal waveguide, which is partially filled with magnetized plasma versus effective wave number. The *numbers* indicate the azimuthal mode numbers m

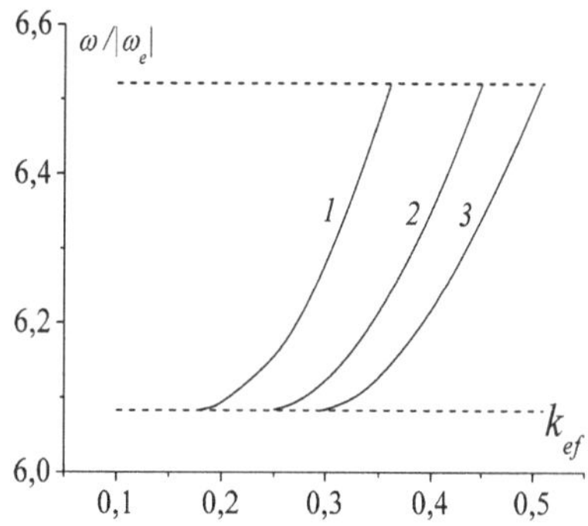


**Fig. 2.3** Frequency of HF SFXMs propagating in a metal waveguide, which is partially filled with magnetized plasma versus effective wave number. The *numbers* indicate the absolute value of the azimuthal mode number |m|

X-modes. Analyzing the analytical expressions derived for LF flute X-modes, one can make the following conclusions:

(1) their dispersion is normal;
(2) their frequency increases with increasing plasma density and vacuum layer thickness.

Figure 2.2 demonstrates the dependence of the LF SFXM frequency normalized to the electron cyclotron frequency: $\omega/\left|\omega_e\right|$, versus the effective wave number $k_{ef} = |m|\,\delta/R_1$ and $\delta = c/\Omega_e$ is the skin depth. The following plasma waveguide parameters were chosen for the calculations: $Z \equiv \Omega_e/\left|\omega_e\right| = 10$, $\Delta = 0,2$, $\varepsilon_d = 1$, $m = 1 \div 7$.

Dispersion curves of HF SFXMs are given in Fig. 2.3 by the solid lines. The dashed lines show the boundaries of the HF range. The dependencies of the HF SFXM frequencies normalized to the electron cyclotron frequency are shown in Fig. 2.3 as functions of the effective wave number $k_{ef}$. For these dispersion curves, the following plasma waveguide parameters have been chosen: $Z = 6$, $\Delta = 0,2$, $\varepsilon_d = 1$. In the limiting case of large values of the radius $R_2$ of the metal waveguide metal the dispersion curves of these SFXMs with different azimuthal numbers differ from each other only slightly; they are confined near the lower boundary of the HF range (2.4). The SFXM dispersion curves are restricted as well by the upper boundary of the HF range.

The results of the numerical analysis of Eq. (2.5) show us that increasing the radius $R_1$ of the plasma column, the external magnetic field $B_0$ and the thickness $\Delta$ of the vacuum layer leads to decreasing frequency of the HF flute X-modes. In contrast to this the frequencies of LF flute X-modes depend on these parameters in a different way: their frequency increases with increasing $\Delta$ and with decreasing $R_1$ and $B_0$ values. Changing the sign of the azimuthal mode number makes a substantial effect on the SFXM dispersion properties. The HF branch of these X-modes vanishes for the case $m > 0$, and the frequency of their LF branch with $m > 0$ considerably grows up in comparison with the corresponding values of their frequencies in the case $m < 0$. Consequently, eigen flute X-modes in such plasma structures are non-mutual: the frequencies of the waves which are characterized by azimuthal mode numbers with identical absolute magnitude, but with opposite signs are different under the same other conditions. This is typical for SWs propagating across an external magnetic field oriented along the plasma boundary [5–7].

Typical radial distributions of LF SFXM fields are shown in Fig. 2.4. The axial magnetic field $H_z$ is continuous and is not equal to zero on the metal wall, $r = R_2$. The radial electric field of LF SFXMs $E_r$ is discontinuous on the plasma-dielectric interface; however the azimuthal electric field $E_\varphi$ is a continuous function of the radial co-ordinate and is equal to zero on the metal wall. The following plasma waveguide parameters were chosen for the calculations: azimuthal mode number $m = -2, Z = 10, \Delta = 0.11, \varepsilon_d = 1, k_{ef} = 0.3, \omega = 1.8831\,|\omega_e|$.

Studying the results of the numerical analysis of Eq. (2.6) allows one to find out that SFWs can propagate as well in the case of an isotropic plasma. In this case, their frequency decreases with increasing radius of the plasma column and decreasing thickness of the dielectric layer. The graphic dependence of their frequency upon the plasma column radius is similar to the hyperbolic one. This also proves that if there is no external magnetic field the SFW frequency depends strongly on the azimuthal mode number, thickness of dielectric layer and radius of the plasma column. In this case the considered modes can be used in antenna devices of plasma electronics [8, 9]. Radiation of their energy can be carried out through dielectric "windows" in the metal wall of the waveguide. Because in this case the wave length of the SFW radiation is determined by the relation $\lambda \approx 6 \cdot 10^6 n_{pl}^{-1/2}$cm, then for the plasma density $n_{pl} \sim 10^{12}$ cm$^{-3}$ one can obtain radiation in the centimeter range, which could be gradually controlled.

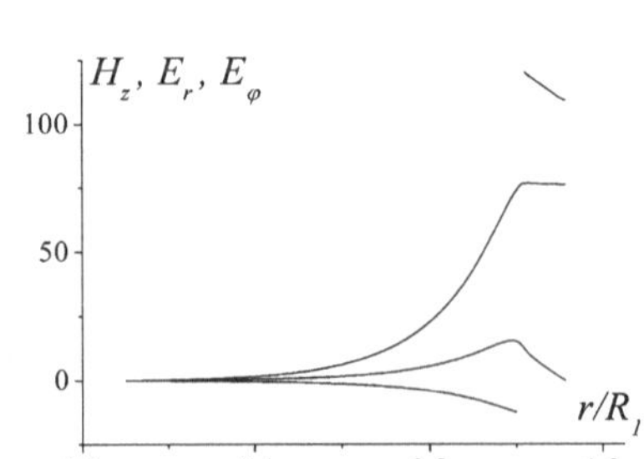


**Fig. 2.4** Radial distribution of the fields of LF SFXMs in the case of propagation along a plasma-dielectric interface

## 2.2 Extraordinary Polarized Surface Flute Waves Propagating along Magneto-Active Plasma-Metal Boundary

Let us consider the properties of SFWs, which propagate along the interface between a uniform cold magneto-active plasma and the metal wall [10] of the corresponding waveguide. Such structures can be realized for the following cases: the cylindrical metal waveguide is entirely filled with plasma or a cylindrical metal rod is immersed into the plasma. Thus one can apply a co-ordinate system, which is the same, as in the previous section, but the radii of the metal and plasma surfaces are equal: $R_1 = R_2 = R$. The other initial assumptions, which have been chosen in Sect. 2.1, are also valid here. That is why the radial distribution of the SFW fields in the plasma region is described in general by expression (2.1). But one can understand that in the case of a plasma cylinder which is surrounded by a metal wall with ideal conductivity the second constant of integration is equal to zero: $A_2 = 0$, and in the case of a cylindrical metal rod which is immersed into magneto-active plasma the constant $A_1 = 0$. Starting from Maxwell equations, one can derive equations, which establish connections between the magnetic and electric field components of the flute X-modes:

$$E_r = \frac{\omega}{ck_\perp^2}\left[\mu\frac{\partial H_z}{\partial r} + \frac{m}{r}H_z\right],\ E_\varphi = \frac{i\omega}{ck_\perp^2}\left[\frac{\partial H_z}{\partial r} + \mu\frac{m}{r}H_z\right]. \tag{2.12}$$

As far as the boundary condition $E_\varphi(r = R) = 0$ is valid on the plasma-metal interface, one can derive from Eqs. (2.1) and (2.12) the following dispersion relation for SFXMs propagating along a plasma-metal interface with finite radius of curvature:

$$k_\perp R F_m'(k_\perp R) + m\mu F_m(k_\perp R) = 0. \tag{2.13}$$

The scheme of the cylindrical metal waveguide entirely filled with magneto-active plasma is shown in Fig. 2.5 and the scheme of the waveguide structure

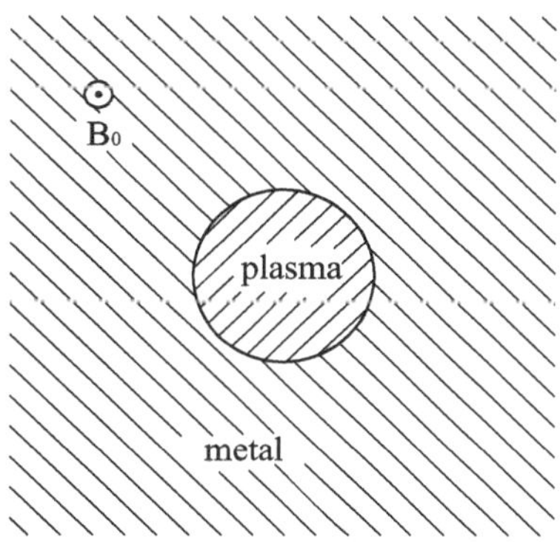


**Fig. 2.5** Scheme of a *metal* waveguide, which is entirely filled with magneto-active plasma

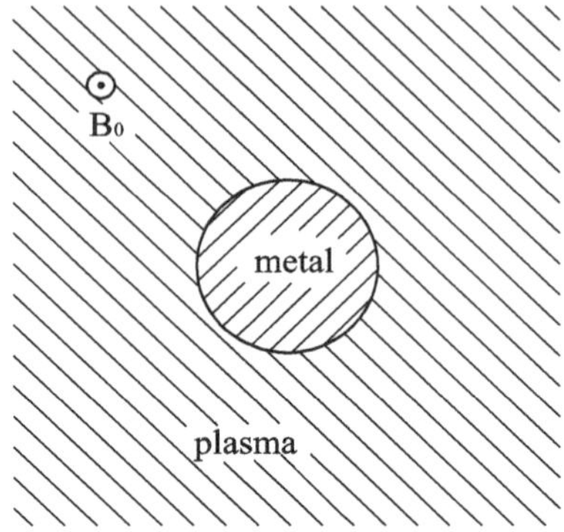


**Fig. 2.6** Scheme of a waveguide structure, which consists of a cylindrical *metal rod* immersed into magneto-active plasma

consisting of a cylindrical metal rod immersed into magneto-active plasma is presented in Fig. 2.6.

In the first case, the function, which is written in (2.13) is as follows: $F_m(\xi) = I_m(\xi)$, and in the second case, this function is of another type: $F_m(\xi) = K_m(\xi)$. In the approximation of large azimuthal mode number: $|m| >> 1$, Eq. (2.13) transforms into the dispersion relation for a SW which propagates exactly across the external magnetic field along a planar plasma-metal boundary [6]. In this case the ratio $m/R$ plays the role of a wave vector. The analysis of the dispersion relation (2.13) proves that in the case under consideration in this section, the frequency range of flute X-modes diminishes as compared with the case of a three-component waveguide structure: metal-dielectric-plasma. Equation (2.13) has solutions in the HF range (2.4) and as well in the LF range, but in contrast to the previous section only in the region $\omega_{LH} < \omega < |\omega_e|$.

Unlike the case that has been investigated in Sect. 2.1, SFWs propagate along the plasma-metal boundary across the axial external constant magnetic field only in one direction in all possible frequency ranges (in other words they are unidirectional waves). In the fluid approximation only X-polarized modes can propagate along the interface between uniform plasma and metal. This corresponds to the general conception of SW theory, in the case of their propagation along a plasma-metal interface [6] across the external magnetic field. The sign of the azimuthal mode number $m$ correlates uniquely with the type of plasma surface curvature. Namely, in the case $F_m = I_m$, the azimuthal mode number is positive: $m > 0$, in the frequency range $\omega < |\omega_e|$, and in the range (2.4) the SFW azimuthal mode number is negative: $m < 0$. For the case of a cylindrical metal rod which is immersed into the plasma, the situation is opposite: $m < 0$ for LF SFXMs ($\omega < |\omega_e|$) and $m > 0$ for HF SFXMs ($\omega_{UH} < \omega$).

Let us derive analytical solutions of dispersion relation (2.13) in some limiting cases. An axial-symmetric ($m = 0$) solution of Eq. (2.13) is realized under the condition $\varepsilon_1^2 = \varepsilon_2^2$. It describes surface oscillations of the plasma cylinder at the cut-off frequencies of bulk modes $\omega = \omega_{1,2}$. In the case $|m| \geq 1$ in the LF range,

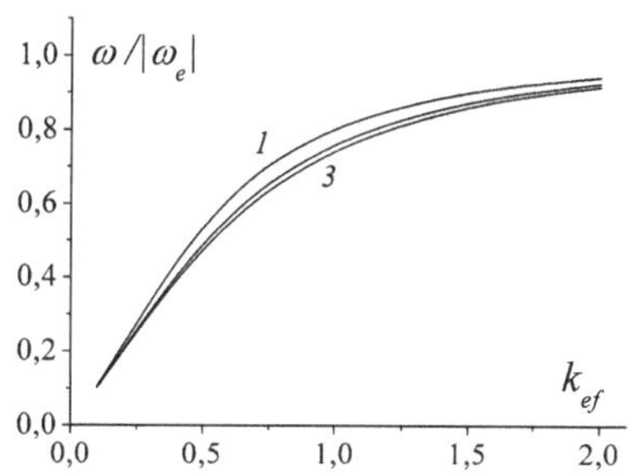


**Fig. 2.7** Dispersion curves of LF SFXMs propagating along a plasma-metal interface for: $\Omega_e = 10\,|\omega_e|$. The *numbers* indicate the absolute value of the azimuthal mode number $|m|$

the spectrum of extraordinary polarized SFWs can be represented by the following expressions:

$$\omega \approx \begin{cases} |m\omega_e|\,\delta R^{-1},\, R > |m|\,\delta, \quad \delta = c/\Omega_e, \\ |\omega_e|\left(1 + R^2\omega_1^2 m^{-2} c^{-2}\right)^{-1/2}, \quad |m| >> 1, \\ |\omega_e|\left[1 + 0.5R^2\delta^{-2}\,(|m|)^{-1}\,(|m| \pm 1)^{-1}\right]^{-1}, \quad R < \delta. \end{cases} \tag{2.14}$$

Here the upper sign relates to the case $F_m = I_m$, and the lower sign to the case $F_m = K_m$. In the case of the HF range, dispersion relation (2.13) has the following solution:

$$\omega \approx \sqrt{\omega_1^2 + \frac{m^2 c^2}{R^2}}, \tag{2.15}$$

for the inequality $\frac{R^2|\omega_e|}{\delta^2\Omega_e} >> m^2 >> 1$.

The frequencies of HF and LF SFXMs propagating along a plasma-metal boundary increase with increasing value of the azimuthal mode number $|m|$ and decreasing value of the waveguide radius $R$. They tend to the maximum values of the LF and HF ranges with increasing $B_0$ value. The difference between SFXMs frequencies in the cases $F_m = K_m$ and $F_m = I_m$ is negligibly small.

Figures 2.7 and 2.8 show the results of a numerical analysis of the dispersion relation (2.13) in the LF and HF ranges, respectively, for the lowest values of the azimuthal mode number $|m| = 1,2,3$. The ratio $\omega/|\omega_e|$ is chosen as ordinate axis. Comparison of these curves and those, which are obtained for the case of waveguides which are partially filled with magneto-active plasma (the last ones are plotted in Fig. 2.2 and Fig. 2.3 for the LF and HF ranges, respectively) allows us to make the following conclusions. First, the dependence of the SFXM frequencies on the absolute value of the azimuthal mode number is absolutely opposite in the cases when these waves propagate along plasma-metal or plasma-dielectric interfaces, respectively. Second, in the case of waveguides entirely filled with plasma the LF range becomes narrow, and the upper range limit becomes $|\omega_e|$ unlike in the case of waveguides only partially filled with plasma. The range of effective wave numbers where the dispersion equation (2.13) has solutions for the HF range becomes wider than that one, which characterizes the dispersion equation of

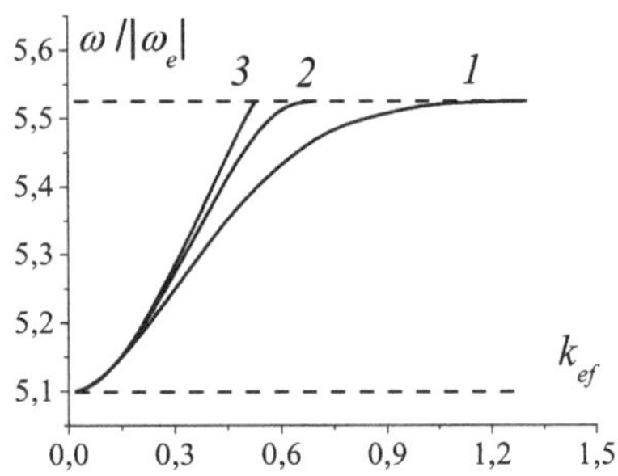


**Fig. 2.8** Dispersion curves of HF SFXMs propagating along a plasma-metal interface for: $\Omega_e = 5\,|\omega_e|$. The *numbers* indicate the absolute value of the azimuthal mode number |m|

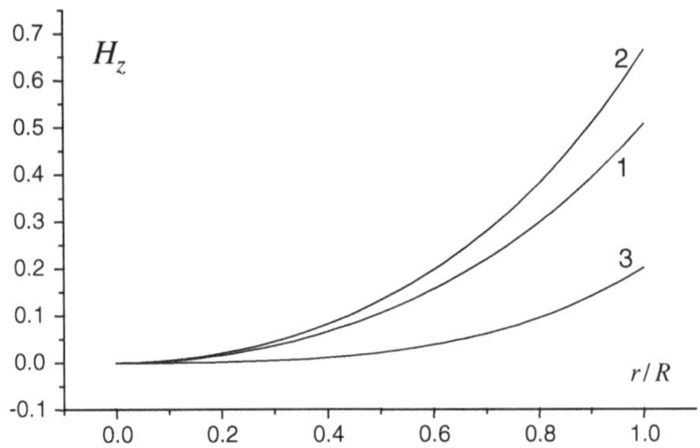


**Fig. 2.9** Radial distribution of $H_z$ for LF SFWs propagating along a plasma-metal interface for the case $R = 2 \cdot \delta$. The numbers *1*, *2* and *3* mark the curves obtained for: $m = 2, \Omega_e^2 = 7\omega_e^2$; $m = 2, \Omega_e^2 = 50\omega_e^2$; and $m = 3, \Omega_e^2 = 50\omega_e^2$, respectively

SFXMs propagating along a plasma-dielectric interface (compare, please, Fig. 2.8 with Fig. 2.3).

As one can see, the SW features, in the case of finite value of curvature radius of the plasma interface, are changed as compared with the case of planar plasma-metal [6] structures. One can point out the following differences. First, the flute X-mode frequency has two limiting values in both the HF and LF ranges. Second, a unique correspondence exists between the sign of the azimuthal mode number and the sign of curvature radius of the plasma surface. Third, the spatial distributions of these transverse surface modes fields propagating along planar and curved plasma-metal interfaces are different, namely: SFWs are extraordinary polarized waves, which have two electrical components and one magnetic component, while in the case of a planar waveguide the tangential electric field of the SW vanishes in the whole plasma volume.

The analysis of Eq. (2.13) proves that SFXMs cannot propagate along a plasma-metal interface in the case of isotropic uniform cold plasma unlike this type of flute waves considered in the previous section. Results of numerical analysis of the radial distribution of the axial magnetic field and azimuthal electric field of SFXMs propagating along a plasma-metal interface are presented in Figs. 2.9 and 2.10, respectively. The radial distribution of the $E_r$ component is similar to the distribution of the $H_z$ component and is therefore not shown. But it should be mentioned that its

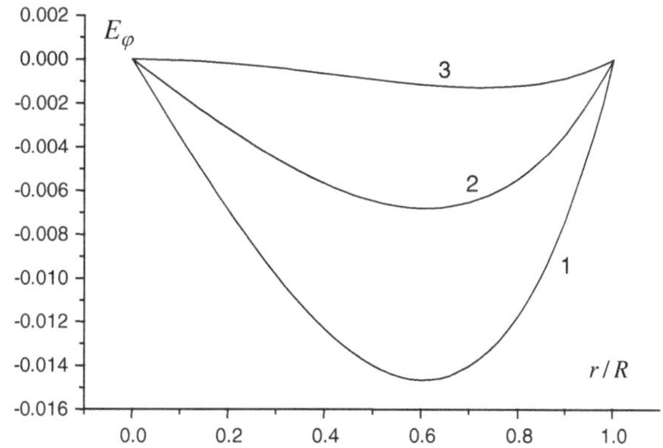


**Fig. 2.10** Radial distribution of $E_\varphi$ for the same cases as in Fig. 2.9

magnitude is six times smaller than the magnitude of the axial magnetic field and its relative contribution to the complete field of SFXMs decreases with increasing plasma density unlike the contribution of the $H_z$ component. The radial distribution of the $E_\varphi$ component of SFXMs is characterized by minimal values of magnitude of this component and the maximum of the $|E_\varphi|$ is located nearby the plasma-metal interface, unlike in the case of plane plasma-metal boundary [6].

## 2.3 Surface Flute Waves Propagating in Strongly Magnetized Plasma Waveguides

Up to now only the case of a dense plasma has been examined ($\Omega_e > |\omega_e|$). Thus the frequency of the electromagnetic perturbations considered above belongs to the range which is higher than the lower hybrid resonance. That is why the motion of plasma ions was neglected there. The indicated correlation between electron plasma and electron cyclotron frequencies can be easily satisfied, for instance in plasmas of semiconductors [5, 11–13] because they are characterized by high concentration of free electrons. However, to produce plasmas under laboratory conditions one can utilize a strong axial magnetic field, so that the other inequality $\Omega_e^2 < \omega_e^2$ can be realized. Thus let us consider just this case [14] when propagation of SFWs is connected with perturbations of plasma ions' motion in waveguides of cylindrical geometry. The scheme of the waveguide considered in this section is presented in Fig. 2.1 as well.

Due to the plasma ions' motion the expressions for the frequency ranges where extraordinarily polarized SFWs can exist, changes compared to the case of application of a weak external magnetic field:

$$\omega_{LH} < \omega < \omega_1 \text{ and } \omega_{UH} < \omega < \omega_2, \tag{2.16}$$

which in this section are called the LF and HF ranges, respectively. Here $\omega_{1,2} = \pm 0.5\,|\omega_e| + \sqrt{0.25\omega_e^2 + \omega_i\,|\omega_e| + \Omega_e^2}$ are the cut-off frequencies of the bulk waves [3, 11] in the case of a strongly magnetized plasma. In this case the HF range becomes very narrow and is located near-by the electron cyclotron frequency, and therefore SFW frequencies in this range depend very weakly on the plasma parameters. Under these conditions the lower hybrid resonance is located in the range of the ion cyclotron frequency $\omega_i$ and therefore there exists the possibility for propagation of plasma ion perturbations across an external magnetic field.

Applying the boundary conditions mentioned above one can derive the dispersion relation for extraordinary polarized SFWs in strongly magnetized plasmas in the form of Eq. (2.5). Analytically it can be examined in the limiting case of a narrow dielectric layer ($m\Delta << 1$,$(R_2 - R_1)k << |m|$), when the influence of the plasma parameters on the dispersion properties of SFWs is the strongest. It is found that unlike in the case of SFWs propagating in waveguides which are partly filled with a weakly magnetized dense plasma, extraordinary polarized SFWs cannot propagate in wide cylindrical metal waveguides with thin dielectric coating which are filled by a strongly magnetized plasma. Mathematically this feature can be proved by analysis of both parts of Eq. (2.6). In the case of strongly magnetized plasmas the left hand side of Eq. (2.6) is positive and the right hand side is negative, therefore the dispersion relation has no solutions [14].

So let us consider the limiting case of a narrow waveguide, which is partly filled by a strongly magnetized plasma: $k_\perp R_1 \ll 1$. Then an analytical expression for the frequency of the flute X-mode can be written in the following form:

$$\omega \approx \omega_i \left[ 1 + \left| \frac{R_2^{2m} - R_1^{2m}}{R_2^{2m} + R_1^{2m}} \right| \Omega_i^2 \omega_i^{-2} \left( \left| \frac{R_2^{2m} - R_1^{2m}}{R_2^{2m} + R_1^{2m}} \right| + \varepsilon_d \right)^{-1} \right], \tag{2.17}$$

here $\Omega_i$ is ion plasma frequency. In this case SFXMs have a unique direction of propagation, namely the azimuthal mode number can be only negative. Thus they are unidirectional waves in this case and the direction of their propagation coincides with the direction of cyclotron rotation of positively charged ions nearby the plasma surface. Thus one can see from Eq. (2.17) that the ion SFW frequency is approximately equal to the ion cyclotron frequency.

Figure 2.11 shows the dependence of the frequency of ion SFWs normalized to the ion cyclotron frequency versus the combination $\Omega_i/\omega_i$ for three values of the azimuthal mode number $m = -2,-3,-4$. Just this combination is chosen as abscissa since the analytical solution shows that the eigen frequency depends mostly on this parameter. The following neon plasma waveguide parameters were chosen for this numerical analysis: $\Delta = 0.1$, $\varepsilon_d = 1$, $|m|\, c/(R_1\Omega_i) = 4.8$. The curves end for small values of abscissa $\Omega_i/\omega_i$ since the frequency reach the lower boundary of the frequency range. The dispersion curve for the case $m = -1$ is not shown in the plot since these ion flute X-modes under the chosen conditions are of bulk nature, so $k_\perp^2 < 0$.

An analysis of the radial distribution of the ion SFXM fields proves that in this case SFWs can be considered as quasi-electrostatic (or in other words, quasi-potential) waves; since the magnetic field of these SFXMs is much less than their

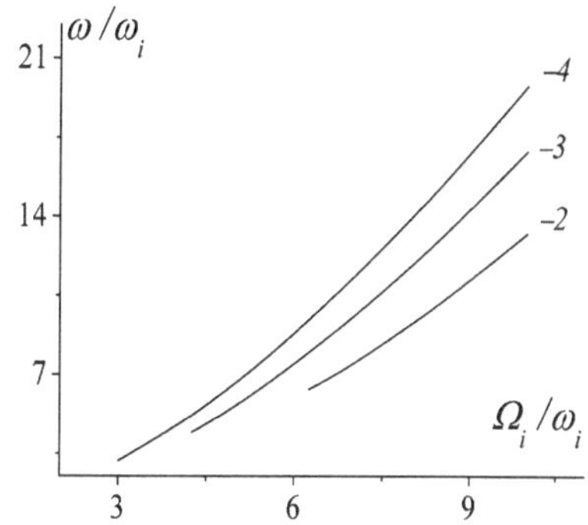


**Fig. 2.11** Frequency of ion SFXMs versus the ratio $\Omega_i/\omega_i$. The *numbers* indicate the azimuthal mode numbers $m$

electric fields. The amplitude of the SFXM electric fields with azimuthal index $m = -1$ remains practically independent on the radial co-ordinate even in the plasma region.

In waveguides with a thick dielectric layer, the properties of flute X-modes depend weakly on the plasma parameters and are determined mainly by the properties of the dielectric layer that separates the plasma column from the metal wall of the waveguide. Thus for this case high radial modes of the SFW can propagate, and a solution of dispersion equation (2.5) exists if the following inequality is valid:

$$(n + 0.5)\pi < (R_2 - R_1)k < (n + 1)\pi; \quad n = 1, 2, 3 \ldots; \quad kR_1 >> |m|. \quad (2.18)$$

In this case SFWs become non-potential waves (their magnetic component is much larger than their electric component) unlike in the previous case of strongly magnetized waveguides with thin dielectric layer. In wide waveguides, which are immersed into strong magnetic fields, SFWs propagate if the inequality $k_\perp R_1 >> |m|$ is valid; the penetration depth of the SFW into the plasma is small and an analytical expression for the angular SFW frequency can be represented in the form $\omega = \omega_{0n} + \delta\omega$, where $\delta\omega$ is their frequency correction; the frequency of their $n$th radial mode is approximately equal to $\omega_{0n} \approx 0.5\pi\ n\ c/(R_2 - R_1)$.

Analyzing the dispersion relation (2.5) in this limiting case, one can make the following conclusions on SFW propagating in a strongly magnetized waveguide, which is partially filled with a cold uniform plasma:

(1) The dependence of the SFW frequency on the azimuthal mode number is very weak in this limiting case;
(2) This dependence becomes weaker and weaker with increasing value of plasma density;
(3) The value of the SFW frequency correction $\delta\omega(m)$ is larger for positive mode numbers $m$ than for negative $m$;
(4) The difference between values of the SFW frequencies with mutually opposite values of azimuthal mode number is maximal nearby the upper band of the LF range (2.16), where the equality $\varepsilon_1 + \varepsilon_2 = 0$ is valid [14].

Results of numerical analysis prove that the frequency of the mode, which has zero radial index, lies approximately in the range between the 20-th and 40-th harmonics of the ion cyclotron frequencies. The limiting values of the frequency range of the first radial mode are approximately equal to 60 $\omega_i$ and 80 $\omega_i$.

As an example let us calculate the eigen frequency of the SFW with $m = -3$ which propagates in a fully ionized helium plasma under the following conditions: plasma radius $R_1 =$ 10 cm, an external magnetic field $B_0 = 0.3\,T$, temperature $T = 1$ eV, plasma density $n_e = 10^{10}$ cm$^{-3}$, parameter $\Delta = 0.1$. Under these conditions the angular frequency of the flute X-mode is as follows: $\omega \approx 19\omega_i = 1.4 \times 10^8 \mathrm{s}^{-1}$.

## 2.4 Ordinary Polarized Flute Electromagnetic Waves

As it was indicated at the beginning of this Chapter, flute electromagnetic perturbations with ordinary polarization can also propagate in cylindrical waveguides filled with magneto-active plasma (see, please Fig. 2.1). In the limiting case of uniform plasma filled waveguides, these O-modes are not interesting from the practical point of view in comparison with the extraordinarily polarized SFWs, because they cannot propagate along the plasma-metal boundary and their properties are mostly determined by the properties of the dielectric layer that separates the plasma column from the metal wall of the waveguide [7]. But in current-carrying plasma filled waveguides SFWs propagate as coupled X- and O- modes. Therefore let us consider the dispersion properties of the ordinary polarized SFWs (SFOMs) propagating in waveguides filled with uniform plasmas using fluid approximation.

In the general case the axial component of the electric field of flute O-modes is described by the following inhomogeneous differential equation:

$$\frac{1}{r}\frac{d}{dr}\left(r\frac{dE_z^{(m)}}{dr}\right) + \left(k^2\varepsilon_3 - \frac{m^2}{r^2}\right)E_z^{(m)} = \widehat{M}H_z^{(m)}, \tag{2.19}$$

where $\varepsilon_3 \approx 1 - \Omega_e^2(r)\omega^{-2}$; the operator $\widehat{M}H_z^{(m)} = 0$ if there is no external azimuthal magnetic field: $B_{0\varphi} = 0$. In the other cases when an external azimuthal magnetic field is applied, $B_{0\varphi} \neq 0$, the explicit form of the operator $\widehat{M}H_z^{(m)}$ is given in Chap. 3 for each specific case. Regardless of the $B_{0z}$ and $B_{0\varphi}$ values of the external constant magnetic field, two other components of the electromagnetic field of flute O-modes are expressed through $E_z^{(m)}$ in the following way:

$$H_r^{(m)} = mE_z^{(m)}(kr)^{-1}, \quad H_\varphi^{(m)} = -k^{-1}dE_z^{(m)}/dr. \tag{2.20}$$

A solution of Eq. (2.19) in a dielectric region, which satisfies the boundary condition for the waves' tangential electric field on the metal wall of the waveguide, namely: $E_z^{(m)}(R_2) = 0$, can be written in terms of Bessel and Neumann functions

[2]. Application of the boundary conditions on the plasma-dielectric interface, which determine continuity of the tangential (in respect to the plasma interface) wave fields allows one to derive the dispersion relation in the form $D_0 = 0$, where:

$$D_0 = \frac{\varphi_0}{d\varphi_0/dR_1} - \frac{1}{k}\frac{J_m(kR_1)N_m(kR_2) - J_m(kR_2)N_m(kR_1)}{J'_m(kR_1)N_m(kR_2) - J_m(kR_2)N'_m(kR_1)}. \tag{2.21}$$

In expression (2.21), $\varphi_0(r)$ is a solution of Eq. (2.19), which is restricted on the waveguide axis, $r = 0$. Analyzing the Eqs. (2.19) and (2.21), one can make the conclusion that the properties of flute O-modes do not depend on the value of an external axial magnetic field. This is typical for all ordinary polarized modes [1, 3, 11, 15].

In the case of a cylindrical metal waveguide which is partially filled by a uniform plasma column, flute O-modes can propagate in the form of ordinary polarized SFWs with frequencies: $\omega < \Omega_e$. In this case, the solution of Eq. (2.19) looks as follows: $\varphi_0 = I_m(kr\sqrt{-\varepsilon_3})$ [7].

SFOMs cannot propagate at any frequency, which is higher than the electron plasma frequency, because ordinary polarized electromagnetic flute modes acquire bulk nature in this frequency range. So the solution of Eq. (2.19) gets the following form: $\varphi_0 = J_m(kr\sqrt{\varepsilon_3})$, where the argument of the first kind Bessel function is real, because $\varepsilon_3 > 0$. If the thickness of the dielectric coating (to simplify our consideration, let us assume that the dielectric constant is equal to unit) is sufficiently small, which means that the inequality $k(R_2 - R_1) << 1$ is valid, the dispersion relation of SFOMs can be reduced and written in the following approximate form:

$$\frac{\varphi_0}{d\varphi_0/dR_1} \approx -(R_2 - R_1). \tag{2.22}$$

From analysis of Eq. (2.22), it becomes clear that SFOMs cannot propagate in this limiting case, because the inequality $d(\ln|\varphi_0|)/dR_1 < 0$ can be valid only for some bulk flute O-modes.

In the opposite limiting case, namely, for waveguides with a thick dielectric layer: $k(R_2 - R_1) >> |m|$, an analytical solution of dispersion relation (2.22) can be found by the method of successive approximations. Then the analytical expression for the SFOMs frequency has the form of a sum $\omega^{(0)} + \omega^{(1)}$, where $\omega^{(0)}$ is the wave angular frequency calculated in zero approximation and $\omega^{(1)}$ is a small correction to it. Since SFOMs propagate within the frequency range $\omega < \Omega_e$ (in this case $\varepsilon_3 < 0$) in the limiting case of large radius of the plasma column ($kR_1 >> |m|$), one can obtain the following expressions for $\omega^{(0)}$ and $\omega^{(1)}$ [7]:

$$\omega^{(0)} = \frac{\pi n c}{2(R_2 - R_1)}\left[1 + \sqrt{1 + \frac{2m^2(R_2 - R_1)^2}{\pi^2 n^2 R_2 R_1}}\right], \quad n = 1, 2, 3\ldots \tag{2.23}$$

$$\omega^{(1)} = \frac{\omega^{(0)}\delta_o}{R_2 - R_1}\left(1 - \frac{m^2}{2k^2R_1^2}\right)\left[1 + \frac{m^2c^2}{2\left(\omega^{(0)}\right)^2 R_2R_1}\right]^{-1}. \tag{2.24}$$

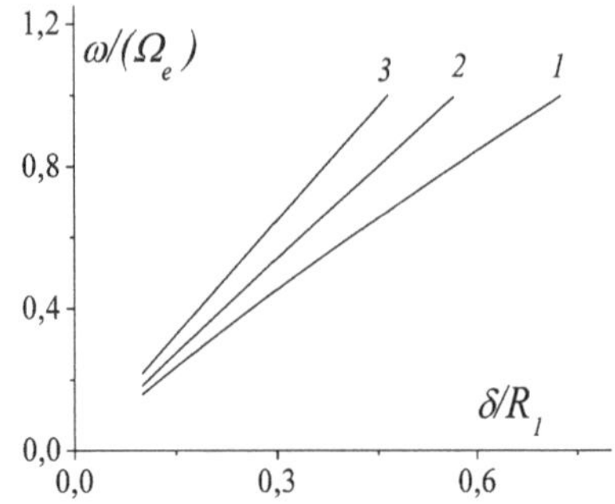


**Fig. 2.12** Frequency of the first radial SFOMs ($n = 1$) versus skin depth normalized to the plasma radius. $R_2 = 3R_1$, $\varepsilon_d = 1$, $m = 1, 2, 3$

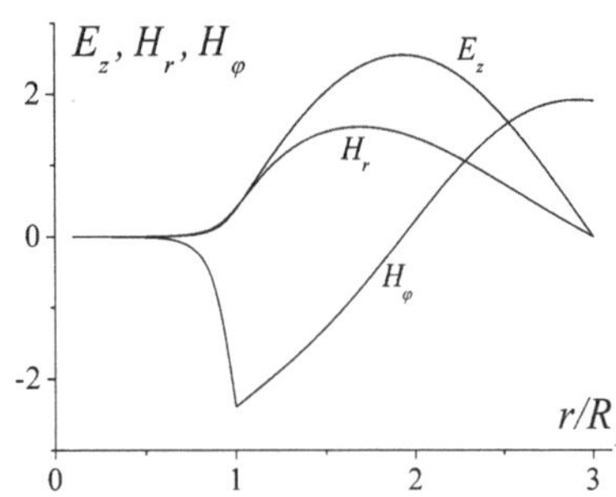


**Fig. 2.13** Radial distribution of the fields of the SFOM with radial mode number $n = 1$, azimuthal mode number $m = 2$ propagating in a waveguide with $\delta = 0.15R_1$, $R_2 = 3R_1$, $\varepsilon_d = 1$

The influence of the plasma density on the properties of SFOMs becomes weaker with decreasing plasma column radius, i.e. if inequality $kR_1 < 1$ is valid:

$$\omega^{(0)} = \frac{\pi nc}{R_2}\sqrt{\left(1 + \frac{|m|}{2n} - \frac{1}{4n}\right)^2 - \left(\frac{m}{\pi n}\right)^2}, \quad n = 1, 2, 3\ldots \tag{2.25}$$

$$\omega^{(1)} = \frac{\pi\,|m|\,c}{R_2\,(|m|!)^2}\left(\frac{kR_1}{2}\right)^{2|m|}\left(1 - \frac{|m|\,\delta_o}{R_1}\right). \tag{2.26}$$

Figure 2.12 presents the results of the numerical analysis of dispersion relation (2.21). It shows the frequency of the first radial ($n = 1$) SFOMs with azimuthal mode numbers $m = 1, 2, 3$ normalized to the electron plasma frequency versus skin depth normalized to the plasma radius. The following plasma waveguide parameters were chosen for these calculations: $R_2 = 3R_1$, $\varepsilon_d = 1$. The top the curves are limited by the boundary of the frequency range $\omega = \Omega_e$.

Radial distributions of the fields of the SFOM which propagates with azimuthal mode number $m = 2$ in the case $\delta = 0.15R_1$ are given in Fig. 2.13. The plasma waveguide parameters, which were chosen for this calculation, are the same as for the previous figure. The axial electric field $E_z$ of the wave is equal to zero on the metal as predicted by the general theory of SWs [5, 6, 16, 17]. The radial magnetic field $H_r$ of the wave is almost proportional to $E_z$, the weak difference is explained by the factor $1/r$ [see Eq. (2.20)]. The azimuthal magnetic field $H_\varphi$ of the wave differs from zero at the metal wall of the waveguide.

Thus the dispersion properties of SFOMs in cylindrical metal waveguides, partially filled with a uniform plasma are determined mainly by the geometry of the utilized waveguide, first of all by the thickness of the dielectric layer (in the dielectric layer they have features of bulk waves). The frequency of SFOMs increases with decreasing dielectric layer thickness and with increasing azimuthal ($m$) and radial ($n$) indices of these modes. The electron plasma frequency is the upper limit of the SFOM frequency range, therefore it is $\sqrt{2}$ times larger than the maximum frequency of SWs propagating along a planar plasma interface [5, 11]. In the frequency range $\omega > \Omega_e$ flute O-modes propagate in the form of bulk waves.

## 2.5 Damping of Surface Flute X-Modes

To calculate the damping rate of surface flute X-modes which is caused by collisions between the plasma particles, one can take into consideration the effective value of the collisional frequency $\nu$ in the components of the dielectric permittivity tensor $\varepsilon_{jk}$ [3, 11, 15] applied in dispersion relation (2.5). This allows one to find the expression for the collisional damping rate of these modes:

$$\gamma_{col} \approx \nu/3. \tag{2.27}$$

Another reason, which can lead to damping the surface flute X-modes is the presence of a slot in the metal wall of the utilized waveguide. Employment of such slotted waveguides allows one to radiate power of the SFXMs out of the waveguides, thus they can be applied as antenna systems [8, 9].

Realization of the electromagnetic power emission out of plasma filled waveguides is an important problem for the development of generators of radio frequency and microwave radiation. The articles [18, 19] are devoted to theory of open periodic plasma waveguides, which are excited by electron beams in the regime of bulk wave generation. In the case of the excitation of bulk waves, the output of microwave power is not a serious technical problem. But in the case of surface wave excitation, one has to solve some difficult problems, e.g. to create and sustain definite profiles of plasma density or magnetic field, or to utilize a metal waveguide with special form, namely its radius can be varied according to a specific law [16, 20].

Let us consider a waveguide whose geometry differs from that one, which is studied in Sect. 2.1 by the presence of a narrow axial slot in the metal chamber of the waveguide. It is oriented along the axial direction and its angular size is small: $2\varphi_0 \ll 2\pi$ (see Fig. 2.14). The plasma density is assumed to be relatively large: $\Omega_e^2 >> \omega_e^2$. Determination of the fields of surface flute X-modes in the dielectric layer allows one to find out that the constant of integration $B_2$ in expression (2.2) has to be calculated applying boundary conditions for the electromagnetic fields on

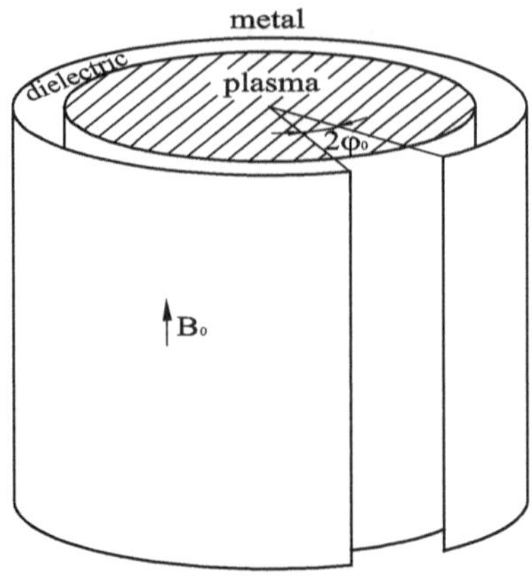


**Fig. 2.14** Scheme of a plasma filled *metal* cylindrical waveguide with a *narrow* axial slot

the interface $r = R_2$, which now have another form as compared with those ones, which were applied in Sect. 2.1. Let us consider these conditions in details here.

They can be determined as follows: first, the azimuthal component of the SFXM electric field is equal to zero on the metal wall of the waveguide [21]:

$$B_2 = -B_1 J_m'(kR_2)/N_m'(kR_2) \quad \text{for } \varphi \in -[\pi, \varphi_0] \cup [\varphi_0, \pi]. \tag{2.28}$$

Second, there is no wave which propagates from the outer space into the inner part of the waveguide through the axial slot in its metal wall:

$$B_2 = iB_1 \quad \text{for } \varphi \in [-\varphi_0, \varphi_0]. \tag{2.29}$$

Since in the case when the slot is absent ($\varphi_0 = 0$), all harmonics of the SFXMs with different values of azimuthal mode number $m$ propagate independently of each other, one can suppose that (in zero approximation) only the fundamental harmonic of the surface flute X-modes with the number $m$ propagate in the waveguide. Then the amplitudes of the $(m \pm 1)$-th, $(m \pm 2)$-th and etc. satellite harmonics have one order smaller amplitudes in comparison with the amplitude of the $m$-th harmonic. Therefore these small-amplitude satellite harmonics make only second order contributions to the dispersion equation of SFXMs [21].

Applying the condition that the SFXM tangential fields are continuous on the plasma-vacuum interface one can derive the dispersion equation for SFXMs in this case. With the accuracy up to terms of first order smallness over the parameter $\varphi_0/\pi << 1$ one can write this equation in the following form [21]:

$$D_0 + D_1 = 0, \tag{2.30}$$

here the expression for the $D_0$ is defined by Eq. (2.5), and the expression for $D_1$ is a small quantity of first order smallness over the angle $\varphi_0$:

$$D_1 = \frac{\varphi_0[iN_m'(kR_2) + J_m'(kR_2)]N_m(kR_1)/\pi}{J_m(kR_1)N_m'(kR_2) - J_m'(kR_2)N_m(kR_1)}\left[\mu\frac{mk}{R_1k_\perp^2} + \frac{kI_m'(k_\perp R_1)}{k_\perp I_m(k_\perp R_1)} + \frac{N_m'(kR_1)}{N_m(kR_1)}\right]. \tag{2.31}$$

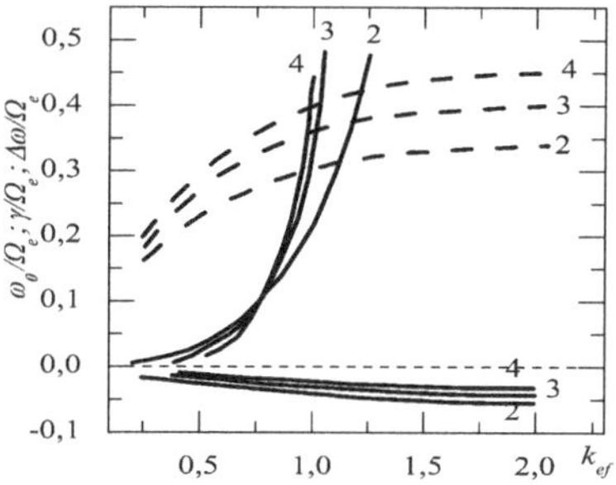


**Fig. 2.15** Eigen frequency of LF SFXMs propagating in waveguide, which is partially filled with plasma (*dashed lines*), their damping rate (*solid line in upper half-space*) and real part of the correction to their eigen frequency (*solid line in lower half-space*) caused by radiation of their power into the outer space versus effective wave number. The *natural numbers* indicate the value of $m$

To solve the dispersion equation (2.30) let us apply the method of successive approximations. Then a solution of the dispersion equation (2.30) can be found as the sum: $\omega = \omega_0 + \Delta\omega - i\gamma$, where $\omega_0$ is the solution of the dispersion equation in the zero ($\varphi_0 = 0$) approximation, $\Delta\omega$ is the real part of the correction to the SFXM eigen frequency $\omega_0$ and $\gamma$ is the damping rate of the SFWs connected with radiation of their power through the axial slot in the metal wall of the waveguide.

In the limiting case of a narrow vacuum layer $\Delta << 1$, for SFXMs with large values of azimuthal mode numbers propagating in wide plasma waveguides $k_0 = m\delta R_1^{-1} << 1$, it is easy to derive an expression of the SFXM damping rate $\gamma$ caused by radiation of power from the waveguide into the outer space [21]:

$$\gamma \approx \Omega_e \varphi_0 (m!)^2 \left(4\delta/m^2 R_1 \Delta\right)^{m+0.5} (4\pi^2)^{-1}. \tag{2.32}$$

The results of numerical analyses of the dispersion relation (2.30) in the LF range are presented in Fig. 2.15. In the upper half-space one can see the frequencies of LF SFXMs (dashed lines) and damping rates caused by radiation of LF SFXM energy into the outer space (solid line) normalized to the electron plasma frequency versus the effective wave vector. In the bottom half-space, the dependence of the real corrections to the frequency is shown. The numbers 2, 3, 4 indicate the azimuthal mode numbers. Following plasma waveguide parameters have been chosen for the calculation: $\Delta = 0.1$; $\Omega_e = 10\,|\omega_e|$; $\varphi_0 = 0.01\pi$.

The results of a numerical analysis of this phenomenon confirm that the SFW frequency $\omega_0$ becomes less than it was in the case of absence of an axial slot and that the additional damping described by $\gamma$ is caused by the leakage of SFXM energy into the outer space. Therefore the presence of an axial slot in the waveguide wall reduces the phase velocity of the SFXMs. Concerning this one can mention the results of paper [18], where such a decrease of the frequency of bulk modes propagating in open periodic waveguide was observed in contrast to the case of closed plasma systems.

The rapid growth of the $\gamma$ value with the increase of $k_{ef} = |m|\, c/(\Omega_e R_1)$ obtained here corresponds to the results which were obtained in paper [19]. But on the other hand, a numerical analysis testifies that the range of existence of SFXMs is located on the left-hand side of the curves $\gamma = \gamma(k_{ef})$, thus in the range of $k_{ef}$ the damping of SFMs caused by radiation of their power through a slot in the wall of the waveguide does not cancel its propagation. As one can see from (2.32) $\gamma \propto \varphi_0/\pi$ (in comparison with the paper [18], where $\gamma \propto a/(a+b)$, here $a$ is the size of the gap in the metal wall, $b$ is the length of the metal sections of open periodic waveguide examined in [18]). However, for efficient operation of an RF generator (oscillator) based on SFW propagation it is required that the SFW damping rate has to be less than the growth rate of their beam instability. This condition is sufficiently satisfied in the range $0.2 \le k_{ef} \le 0.6$. Numerical analysis proves that a reduction of $\gamma$ can be achieved by reducing the magnitude of the magnetic field $\vec{B}_0$ and by increasing the value of the $\Delta$ parameters.

Let us consider the case of SFXMs propagation in a metal waveguide, which is entirely filled with cold magneto-active plasma ($\Delta = 0$) [10]. The equation which describes the radiation of SFXM power through the slot in the metal wall of such a waveguide has the form of Eq. (2.30), where the term $D_1$ has the following form with accuracy up to first order terms of smallness over the parameter $\varphi_0/\pi << 1$:

$$D_1 = \frac{\varphi_0}{2\pi}\frac{J_m(\zeta)J'_m(\zeta) + N_m(\zeta)N'_m(\zeta)}{J_m^2(\zeta) + N_m^2(\zeta)} + \frac{i\varphi_0}{\pi^2 kR}[J_m^2(\zeta) + N_m^2(\zeta)]^{-1}, \quad (2.33)$$

and the expression for the term $D_0$ is as follows:

$$D_0 = \frac{km\mu}{k_\perp^2 R} + \frac{kI'_m(\xi_0)}{k_\perp I_m(\xi_0)}, \quad (2.34)$$

here $\zeta = kR$, $\xi_0 = k_\perp R$.

One can find an analytical expression for the corrections to the frequencies of SFXMs caused by radiation of their power through the slot in the metal wall of the waveguide, which is entirely filled with magneto-active plasma in the limiting cases of narrow ($|m| >> \zeta$) and wide ($|m| << \zeta$) waveguides [10].

Let us write down here the real part of the frequency correction to the SFXM eigen frequency in first approximation over the angular size of the slot in the metal wall of the waveguide, which is caused by radiation of the SFW power into the outer space:

$$\Delta\omega \approx \Omega_e \varphi_0/\pi\varsigma, \quad (2.35)$$

where $\varsigma = 1$ in the case of a narrow waveguide and $\varsigma = 4$ in the case of wide waveguide.

The damping rate of SFXMs caused by their radiation through a narrow axial slot in the metal wall of the waveguide, which is entirely filled with magneto-active plasma, is described by the following expression:

$$\gamma \approx \varphi_0 \Omega_e \varsigma_0/2\pi, \quad (2.36)$$

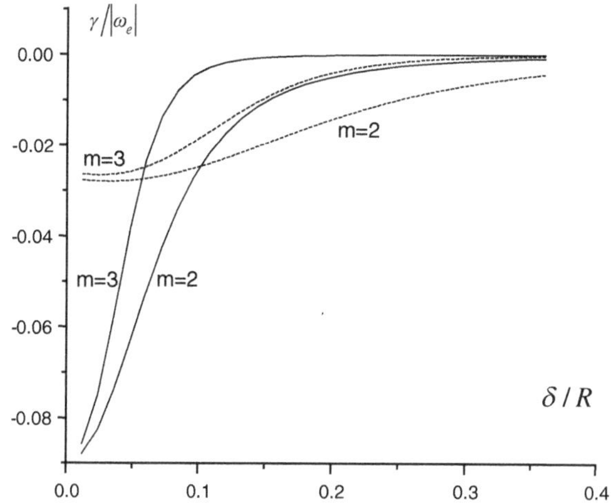


**Fig. 2.16** Dependence of the imaginary part of the frequency correction of the LF SFXMs determined by leakage of wave power from the waveguide, which is entirely filled with magneto-active plasma into the outer space. *Solid* and *dashed lines* are calculated for the cases: $\Omega_e^2 = 50 \cdot \omega_e^2$ and $\Omega_e^2 = 7 \cdot \omega_e^2$, respectively

where $\varsigma_0 = 1$ in the case of wide waveguide and $\varsigma_0 = \frac{4(m!)^2}{\pi|m|}(\zeta/2)^{2|m|}$ in the case of a narrow waveguide.

Figure 2.16 presents results of numerical calculations of the damping rate of LF SFXMs determined by their power radiated through a narrow axial slot in the waveguide wall into the outer space (the angular size of the axial slot was chosen to $\varphi_0 = 0.026 \cdot \pi$). The results of a numerical analysis of this case prove that in a wide waveguide, the value of the real corrections to the LF SFXM frequency increases with increasing value of azimuthal mode number, and for narrow waveguides has the opposite dependence. Increasing the density of the plasma, which is filled into the waveguide, and decreasing the waveguide radius leads to an increase of the value of the SFXM frequency correction $\Delta\omega$. In addition, in the case of $\Delta = 0$, the inequality $\Delta\omega > 0$ is valid. This result differs from that one obtained in [21] for plasma-dielectric-metal waveguide structures, because the existence of an axial slot in a metal wall of the waveguide in the present case (the metal waveguide is assumed to be entirely filled by the plasma) means that the waveguide changes from a two-component plasma-metal structure into a three-component structure: plasma-insulator–metal. As far as it was shown in Sect. 2.2 the frequency of surface flute X-modes is larger just in the case of a three-component waveguide structure that is why such modification of the waveguide has led to an increase of the SFW frequency.

Summarizing the results obtained for the case when the waveguide is entirely filled with a cold magneto-active plasma one can conclude that the damping rate $\gamma$ of surface flute X-modes is determined by the leakage of their power through a narrow axial slot. It increases with decreasing utilized magnetic field, with increasing radius of the waveguide and also with increasing angular dimension of the slot. The value of the real part of the SFXM frequency correction $\Delta\omega$ becomes positive unlike in the case of the presence of a vacuum layer between plasma column and waveguide wall and $\Delta\omega$ decreases with increasing value of the applied magnetic field and with increasing radius of the waveguide.

## 2.6 Long-Wavelength Azimuthally Non-Symmetrical Surface Waves

In all the previous sections of this Chapter we have examined only those waves, which propagate exactly perpendicular to the axis of symmetry of the corresponding waveguide structures. This of course puts some limits on the scope of application of the presented theoretical results, because electromagnetic waves often have a definite axial wave number of a finite value. That is why this section presents the results of studying the properties of extraordinary polarized electromagnetic perturbations, which are characterized not only by the azimuthal mode number $m$, but also by the axial wave number $k_z$. Of course these modes cannot be entitled as flute modes, but nevertheless we shall conduct the study on the base of the theory of surface flute waves (SFW) propagating in plasma filled waveguides. It is quite possible to do that for the case of long-wavelength azimuthally non-symmetrical modes (ANSMs). The problem of studying the dispersion properties of a long-wavelength ANSM can be solved by methods of successive approximation, where the SFW theory is used as zero approximation. Here the analogy between these eigen waves and the bulk resistive drift waves considered in [1], whose propagation is accompanied with a dissipative instability, should be mentioned. But unlike them, long-wavelength ANSMs are stable modes and the form of their dispersion equation is similar to the form of an equation for coupled X- and O-modes.

The presented results are devoted to investigation of the properties of long-wavelength ANSMs propagating in a cylindrical metal waveguide entirely filled with a cold magneto-active plasma, whose density is radially non-uniform [22]. The geometry of the waveguide and the applied assumptions are described in Sect. 2.2, the external magnetic field is directed along the axis of the waveguide: $\vec{B}_0 \parallel \vec{z}$ (see Fig. 2.5). In the case of SFW propagation Maxwell equations split into two independent subsystems, namely for X- and O-modes. But for the long-wavelength ANSMs propagating in radially non-uniform plasma, if one can neglect terms of the second and higher orders of smallness for $k_z$, Maxwell equations can be reduced into two second order differential equations for the axial components of electric and magnetic fields:

$$\frac{1}{r}\frac{\partial}{\partial r}\frac{r}{k_X^2}\frac{\partial H_z}{\partial r} - \left[1 + \frac{m^2}{r^2 k_X^2} - \frac{m}{r}\frac{\partial}{\partial r}\left(\frac{\mu}{k_X^2}\right)\right] H_z = \widehat{K} E_z; \tag{2.37}$$

$$\frac{1}{r}\frac{\partial}{\partial r} r \frac{\partial E_z}{\partial r} - \left(k_0^2 + \frac{m^2}{r^2}\right) E_z = \widehat{M} H_z. \tag{2.38}$$

The small value right-hand parts of the inhomogeneous differential Eqs. (2.37) and (2.38) can be written in the following forms:

$$\widehat{K} E_z = i N_z \left\{ \frac{m}{r} E_z \frac{\partial}{\partial r}\left(\frac{1}{k_X^2}\right) + \frac{\mu k_0^2}{k_X^2} E_z + \frac{\partial}{\partial r}\left(\frac{\mu}{k_X^2}\right)\frac{\partial E_z}{\partial r} \right\}, \tag{2.39}$$

$$\widehat{M}H_z = iN_z \left\{ \frac{1}{N_X^2}\frac{\partial\mu}{\partial r}\frac{\partial H_z}{\partial r} + \left[ \frac{\omega^2}{c^2}\mu + \frac{m}{r}\frac{\partial}{\partial r}\left(\frac{1}{N_X^2}\right) - \frac{\mu m}{r}\frac{\partial}{\partial r}\left(\frac{\mu}{N_X^2}\right) \right] H_z \right\}, \tag{2.40}$$

where $N_X^2 = (\mu^2 - 1)(\varepsilon_1 - N_z^2)$; $N_z = ck_z/\omega$; $\mu = \varepsilon_2/(\varepsilon_1 - N_z^2)$. The penetration depths of the O-mode $k_O^{-1}$ and X-mode $k_X^{-1}$, with field components $(E_z, H_r, H_\varphi)$ and $(E_r,\ E_\varphi, H_z)$, respectively, are determined by the aid of the dielectric permittivity tensor $\varepsilon_{ik}$ (see e.g. [3, 11, 15]). Their explicit forms are as follows:

$$k_0^2 = -\varepsilon_3\omega^2 c^{-2}, \quad k_X^2 = N_X^2\omega^2 c^{-2}. \tag{2.41}$$

The set of Eqs. (2.37) and (2.38) for the longitudinal component $E_z$ of the electromagnetic field in the case of arbitrary values of $k_z$ and radially uniform plasma immersed into an axial steady magnetic field was obtained in [23]. But its analytical solutions for the limiting case of long-wavelength non-symmetrical electromagnetic perturbations were found by two co-authors of the present book in [22] at first time.

The solutions of Eqs. (2.37) and (2.38) have been found by the method of constant variation. The azimuthal component of the ANSM electric field is expressed in terms of $E_z$ according to [24]. Then using the boundary conditions indicated in the Sect. 2.2 for the tangential components of the electric field on the inner surface of the metal waveguide, one can derive the dispersion equation for long-wavelength ANSMs [22]:

$$A_{12}A_{21} = A_{11}A_{22}, \tag{2.42}$$

here

$$A_{21} = \varphi(R_1), \tag{2.43}$$

$$\begin{aligned} A_{11} = &\left\{ \frac{ic}{\omega N_X^2}\frac{\partial\tilde{\psi}}{\partial r}\int\limits_0^{R_1}\frac{\psi k_X^2\widehat{K}\varphi dr}{W\left(\psi,\tilde{\psi}\right)} + \frac{i\mu N_\vartheta}{N_X^2}\tilde{\psi}\int\limits_0^{R_1}\frac{\psi k_X^2\widehat{K}\varphi dr}{W\left(\psi,\tilde{\psi}\right)} \right. \\ &\left. + \frac{N_z N_\vartheta}{N_H^2}\varphi + \frac{\mu N_z c}{\omega N_H^2}\frac{\partial\varphi}{\partial r} \right\}_{|r=R_1}, \end{aligned} \tag{2.44}$$

$$A_{12} = \left\{ \frac{ic}{\omega N_X^2}\frac{\partial\psi}{\partial r} + \frac{\mu N_\vartheta}{N_X^2}\psi \right\}_{|r=R_1}, \quad N_\vartheta = cm/(\omega r), \tag{2.45}$$

$$A_{22} = \tilde{\varphi}\,(R_1)\int\limits_0^{R_1}\frac{\varphi\widehat{M}\psi dr}{W\,(\varphi,\tilde{\varphi})}, \tag{2.46}$$

where $\varphi$ and $\psi$ are solutions of the Eqs. (2.38) and (2.37) respectively, which have no singularities on the axis of the considered waveguide; $\tilde{\varphi}$ and $\tilde{\psi}$ are linearly independent solutions of these equations, which have singularities on the axis of the waveguide.

It should be noted that on the one hand, the dispersion equation (2.42) is a generalization of the dispersion equation derived in [7, 10] for the case of small $k_z$, and on the other hand it is a particular case of the general dispersion equation derived for determination of the frequency range of a coaxial plasma waveguide immersed into a finite constant magnetic field in [23]. The structure of Eq. (2.42) is typical for the case of weakly coupled electromagnetic waves. The condition $A_{12} = 0$ plays the role of the SFXM dispersion equation, whose solution $\omega_0$ is calculated in Sect. 2.2. This means that here we can assume that its value is known. The factor $A_{12}$ in the frequency range found there for the SFXM changes its sign. The condition $A_{21} = 0$ represents the dispersion equation, which describes the properties of bulk flute O-modes.

The right side of the dispersion equation (2.42) is quadratic over the small value $k_z$, it plays the role of a small coupling parameter between the X- and O-modes. Therefore, the frequency correction $\Delta\omega$ to the eigen frequency of long-wavelength ANSMs in the general case is as follows: $\Delta\omega \propto k_z^2$. But this case needs numerical calculation (see [23]) for studying the dispersion properties of these non-symmetrical waves. Unlike that we shall show a possibility to find analytical solutions of their dispersion equation.

Let us find analytical expressions for the solutions of Eq. (2.42) in the limiting case of electromagnetic perturbations with long wavelengths propagating in a plasma with uniform density. In zero approximation the eigen frequency is the same as found in Sect. 2.2. The condition $k_X^2 > 0$ determines the frequency ranges, where ANSMs can exist, which under the condition $k_z = 0$ coincide with the frequency bands where the SFW can exist (see Sect. 2.2). Thus ANSMs with $m > 0$ can propagate in the following frequency range, which will be entitled as low-frequency (LF) range in this section:

$$|\omega_e| \sqrt{(\Omega_i^2 + c^2k_z^2)/(\Omega_e^2 + \omega_e^2)} < \omega < |\omega_e| . \tag{2.47}$$

The ANSMs with negative sign of the azimuthal mode number ($m < 0$) will be entitled here as high frequency (HF) waves, their frequency bands are determined by the following non-equality:

$$\sqrt{\omega_e^2 + \Omega_e^2 + c^2k_z^2} < \omega < 0.5\,|\omega_e| + \sqrt{0{,}25\omega_e^2 + \Omega_e^2 + c^2k_z^2}. \tag{2.48}$$

Thus the studied ANSMs are unidirectional waves, which can propagate only along a single direction in the definite frequency range. This property of SWs is typical for waves propagating across an external constant magnetic field along a plasma-metal boundary (see e.g. [6]). This can be a useful property for wave applications in different electronic devices, where one can avoid propagation of reflected signals. The value of $k_0^2$ has different signs in the ranges (2.47) and (2.48), namely it is positive $k_0^2 > 0$ in the LF range and negative $k_0^2 < 0$ in the HF range.

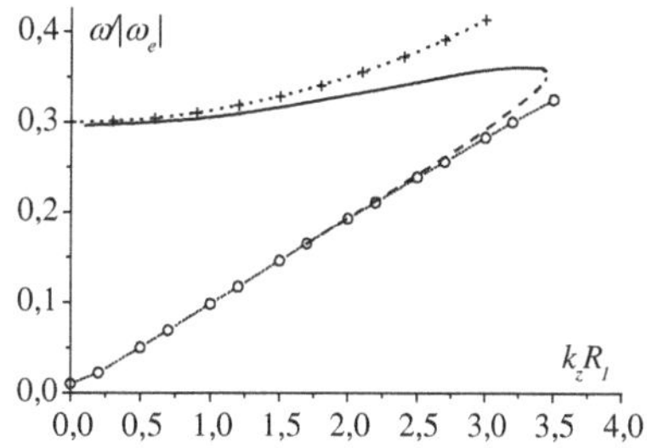


**Fig. 2.17** Normalized frequency of the LF ANSM as function of normalized axial wave number in the case $m = 3$, $\Omega_e = 5\,|\omega_e|$, $R_1 = 10\delta$

In the case of long-wavelength ANSMs propagating in a uniform plasma filled waveguide, dispersion equation (2.42) can be simplified. Its coefficients are given explicitly in [22], so we shall not write down them here. Consideration of small values of $k_z$ leads to the appearance of another additional solution of the dispersion equation (compared with the case of SFWs). In this section, this solution $\omega_-$ will be called as slow mode. For the LF range the $\omega_-$ is almost proportional to the value of $k_z$, thus its value turns to the lower limit of this range at $k_z \to 0$ (see Fig. 2.17). The group velocity of the slow mode is always positive and its value exceeds the absolute value of the group velocity of the second solution of Eq. (2.42), which can be called here as fast mode $\omega_+$. In the limiting case $k_z \to 0$ the fast mode frequency $\omega_+$ becomes a frequency of a SFXM. The dispersion curves of the fast mode have a narrow part of $k_z$ values, where its dispersion is normal. At this part of the ANSM dispersion curve, a small value of $k_z$ leads to an increase of the ANSM frequency compared with the frequency of the SFW calculated for the same conditions. This allows us to calculate the frequency $\omega_+$ of the fast mode by the method of successive approximations: $\omega_+ = \omega_0 + \Delta\omega$, using the theory of SFWs as zero approximation. The small frequency correction $\Delta\omega$ to the frequency of the SFW turns out to be quadratic over $k_z$ in this case,

$$\Delta\omega \approx \left( \frac{A_{11}A_{22}}{A_{21}} - \frac{\partial A_{12}}{\partial k_z^2}\bigg|_{\omega=\omega_0} k_z^2 \right) \left[ \frac{\partial A_{12}}{\partial \omega}\bigg|_{\omega=\omega_0} \right]^{-1} \propto k_z^2. \quad (2.49)$$

A simple analytical form of expression (2.49) can be obtained in two limiting cases of: wide waveguide $R_1 >> |m|\,\delta$, here $\delta$ is a plasma skin depth, and narrow waveguide $R_1 << |m|\,\delta$. Let us write down explicit expressions of the LF ANSM frequencies for the both cases, respectively, in the form of a sum $\omega = \omega_0 + \Delta\omega$:

$$\omega = |\omega_e|\,\frac{m\delta}{R_1} + \frac{3}{8}\,|\omega_e|\,\frac{m\delta}{R_1}\left(\frac{k_zR_1}{m}\right)^2, \quad (2.50)$$

$$\omega = |\omega_e|\left(1 - \frac{R_1^2}{2m^2\delta^2}\right) + \frac{|\omega_e|\,R_1^2}{8\delta^2(m+1)^2}\left(1 - \frac{R_1^2}{2m^2\delta^2}\right)\left(\frac{k_zR_1}{m}\right)^2. \quad (2.51)$$

But numerical analysis has proved that in practice the method of successive approximations can be used to study the dispersion properties of the fast mode of long-wavelength ANSM in the LF range just in a fairly wide range of axial wave numbers $k_z$. In the limiting case of wide waveguides this range is determined by the following inequality:

$$k_z^2 << 8m^2/R_1^2, \tag{2.52}$$

and in the limiting case of narrow waveguides it becomes even somewhat wider:

$$k_z^2 << 8\left(m(m+1)\delta\Big/R_1^2\right)^2. \tag{2.53}$$

The results of the numerical analysis of dispersion equation (2.42) for this case are presented in Fig. 2.17. The frequency of the fast mode of ANSMs at first increases with $k_z$, then it reaches a maximum, and after that it decreases so that the frequencies of the fast and slow modes become close to each other. Finally, for the axial wave number $k_z = k_m$ their frequencies are equal to each other. Nearby this value $k_m$ our representation of eigen electromagnetic perturbations in the harmonic form is incorrect. In [22] the axial width, to which the initial electromagnetic pulse can spread, is estimated. In Fig. 2.17 the solid line shows the dispersion curves of the fast mode and the dashed line indicates the curve of the slow mode for the case of LF ANSMs. The dotted line with hollow circles marks the limit of the LF ANSM frequency band (2.47). The dashed line supplied with crosses is drawn according to the direct calculation of the asymptotic formula (2.50). In the case of a narrow waveguide, the frequency correction is expressed very poorly, so this graph is not shown.

Now let us consider the case of long-wavelength HF ANSMs propagating in a uniform plasma filled waveguide. In the HF range (2.48), a bulk azimuthally non-symmetrical O-mode can also propagate in such waveguides under definite conditions. Thus it can interact with the corresponding X-mode in an ordinary manner and in a resonant way as well. But here we do not consider this problem (see for details [22]). The analysis of dispersion equation (2.42) shows that the HF ANSMs can propagate only in wide waveguides with radius: $R_1/\delta >> \Omega_e\,|m|\,/\,|\omega_e|$ and the bulk azimuthally non-symmetrical O-mode cannot propagate if the external magnetic field and the waveguide radius are not too large, so that the following inequality is valid:

$$R_1^2\omega_e^2\Big/c^2 < j_{|m|,s}^2 - m^2, \tag{2.54}$$

where $j_{|m|,s}$ is the s-th root of the Bessel functions of the first kind $J_{|m|}(j_{|m|,s}) = 0$.

In this case, the HF ANSM frequency and its correction can be found analytically. Their explicit approximate expressions are as follows:

$$\omega_0 \approx \sqrt{\omega_{UH}^2 + m^2c^2/R_1^2}, \tag{2.55}$$

$$\Delta\omega \approx 0.5N_z^2\omega_0 + 0.5\pi\,|m|\,N_z^2J'_m\left(\varsigma^*\right)N_{|m|}(\varsigma^*)\omega_0, \tag{2.56}$$

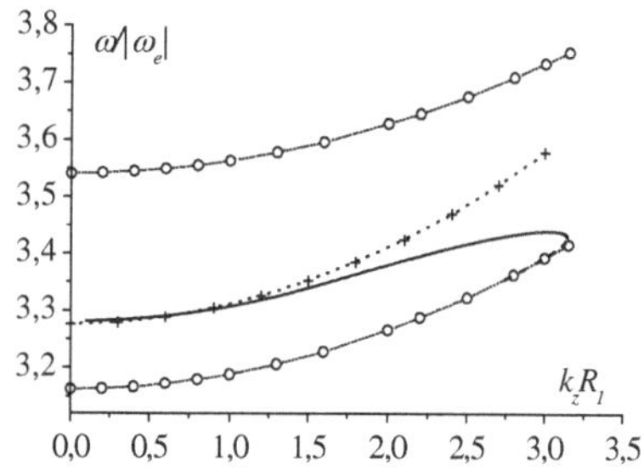


**Fig. 2.18** Normalized frequency of the HF ANSM as function of normalized axial wave number in the case $m = -2$, $\Omega_e = 3\,|\omega_e|$, $R_1 = 7\delta$

where $\varsigma^* \approx \sqrt{m^2 + \omega_e^2 R_1^2 c^{-2}}$. Results of numerical analyses of Eq. (2.42) for the HF ANSM are presented in Fig. 2.18 for the case $m = -2$, with the same notations as in Fig. 2.17. One can see that the width of this frequency range is narrower compared with the case shown in Fig. 2.17. So the slow mode frequency does not become zero when $k_z \to 0$. It should be noted that dispersion curve, which describes the bulk O-mode in this case, lies above the HF range (2.48) where the long-wavelength X-mode can propagate. So there is no resonant interaction between these two ANS modes in this case (see Ref. [22] for details of the resonant interaction).

In this section we have presented results of a theoretical study of the properties of non-symmetric SWs with arbitrary azimuthal mode number and small value of $k_z$, which propagate in a cylindrical metal waveguide entirely filled with magneto-active cold plasma. The dispersion equation for these SWs has been obtained in the case of a non-uniform radial plasma density profile, but its analysis is performed analytically in the limiting case of a uniform plasma. Results of the numerical study confirm the analytical ones.

## 2.7 Conclusions

Results of theoretical investigations on the properties of electromagnetic flute waves propagating in different plasma filled waveguides are presented here for the case of application of an axial magnetic field. The investigations carried out here allow one to determine bands of frequency ranges, within which ordinary and extraordinary flute modes can propagate, to analyze their dispersion properties and radial distribution of their fields, to calculate their damping rates, which are caused by the following physical mechanisms: collisions between plasma particles and radiation of their energy through a slot in the metal wall of the waveguide. It is proved that in the presence of an external axial magnetic field eigen flute waves can propagate both along plasma-dielectric and plasma-metal boundaries and if there is no external magnetic field, SFWs can propagate along a plasma-dielectric boundary only. Phenomena of unidirectional propagation of surface flute X-modes along a plasma-metal boundary have been proved here. From the theoretical point of view such phenomena are typical for surface waves propagating along a plasma-metal interface across an external magnetic field and from practical point of view it is a very useful property in the case when one can obtain a radio-technical device without reflected waves. As already mentioned above, SFW damping can be caused by two reasons, namely by collisions between plasma particles and by radiation of their power through a narrow axial slot in the metal wall of the waveguide. The magnitude of the SFXM damping rates in the both cases (presence and absence of the dielectric layer, which could separate the plasma column from the waveguides wall) is proportional to the angular size of the slot and to the electron plasma frequency. In addition to the surface flute X-modes the same type of O-modes can propagate in waveguides where the plasma column is separated from the waveguide wall by a dielectric layer. The properties of these O-modes are determined by the plasma density and characteristics of the dielectric layer. It is proved as well that application of the SFWs theory allows one to describe dispersion properties even of the eigen non-symmetrical surface waves, which have a small, but a finite value of an axial wave number $k_z$. The values of eigen frequencies of these non-symmetrical waves are found with accuracy up to addendums proportional to the squared value of the $k_z$. The theory of SFWs developed in this Chapter is applied in Chap. 3 for studying the linear interaction between flute X- and O-modes in current-carrying plasma waveguides, in the Chap. 4 for studying SFWs propagating in non-uniform waveguides, and in the Chap. 5 for studying features of the SFWs propagation in waveguides with non-circular cross-section.

## References


1. Chen, F. F. (1974). *Introduction to plasma physics*. NY, London: Plenum Press.
2. Abramowitz, M., & Stegun, I. A. (Eds.). (1964). *Handbook of mathematical functions*. New York: National Bureau of Standards, Applied Mathematics. Series 55.
3. Krall, N. A., & Trivelpiece, A. W. (1973). *Principles of plasma physics*. New York: MacGraw-Hill.
4. Girka, V., Girka, I., Kondratenko, A., et al. (1989). Azimuthal surface waves at the boundary between a magneto-active plasma and a metal. *Soviet Journal of Communications Technology and Electronics, 34*, 96–99.
5. Vukovic, V. (1986). *Surface waves in plasmas and solids*. Singapore: World Scientific.
6. Azarenkov, N. A., & Ostrikov, K. N. (1991). Surface magnetoplasma waves at the interface between a plasma-like medium and a metal in a Voigt geometry. *Physics Reports, 308*, 333–428.
7. Girka, V., Girka, I., Pavlenko, I., et al. (2012). Coupled azimuthal modes propagating in current-carrying plasma waveguides. *Journal of Plasma Physics, 78*(2), 105–123.
8. Anderson, T. (2011). *Plasma antennas*. Boston, London: Artech House.
9. Jenn, D. C. (2003). *Plasma antennas: Survey of techniques and the current state of the art.* SPAWAR PMW 189: San Diego, CA.
10. Girka, V. (2006). Theory of transverse surface electro-magnetic waves propagating along a plasma–metal boundary with a finite radius of curvature in a magnetic field. *Plasma Physics Reports, 32*(5), 401–410.
11. Alexandrov, A. F., Bogdankevich, L. S., & Rukhadze, A. A. (1984). *Principles of plasma electrodynamics*. Berlin and New York: Springer.

12. Beletski, N. N., Bulgakov, A. A., Khankina, S. I., & Yakovenko, V. M. (1984). *Plasma instabilities and non-linear phenomena in semi-conductors*. Kyiv: Naukova dumka. (in Russian).
13. Agranvich, V. M., & Mills, D. L. (1982). *Surface polaritons*. Amsterdam: North-Holland.
14. Girka, I., & Kovtun, P. (1998). Azimuthal surface waves in a magnetized plasma. *Technical Physics, 43*(12), 1424–1427.
15. Akhiezer, A. I., Akhiezer, I. A., Polovin, R. V., Sitenko, A. G., & Stepanov, K. N. (1975). *Plasma electrodynamics*. Oxford: Pergamon Press.
16. Kondratenko, A. M., & Kuklin, V. M. (1988). *Basis of plasma electronics*. Moscow: Energoatomizdat. (in Russian).
17. Aliev, Y. M., Schluter, H., & Shivarova, A. (2000). *Guided-wave-produced plasmas*. New York: Springer.
18. Gestrina, G. N. (1988). Radiation of SHF oscillations of plasma from opened periodic structures. *Soviet Journal Radiophysics and Quantum Electronics, 31*, 1502. (in Russian).
19. Gestrina, G. N. (1989). *On spectrum of eigen frequencies of opened periodic plasma waveguides*. Reports of Ukrainian Academy of Science, *A–10*, 74 (in Russian).
20. Shestopalov, V. P. (1976). *Diffraction electronics*. Kharkiv: Vyscha Shkola. (in Russian).
21. Girka, V. O., & Girka, I. O. (1992). Emission of azimuthal surface waves from narrow waveguide slot. *Soviet Journal of Communications Technology and Electronics, 37*(9), 32–35.
22. Girka, V., & Girka, I. (2002). Asymmetric long-wavelength surface waves in magnetized plasma waveguides entirely filled with plasma. *Plasma Physics Reports, 28*(11), 916–924.
23. Kuzelev, M. V., Romanov, R. V., Selivanov, I. A., et al. (1994). Spectrums of eigen oscillations of coaxial plasma waveguide immersed into finite value magnetic field. *Proceedings of Institute for General Physics RAS, 45*, 17–38 (in Russian).
24. Girka, O. I., Girka, I. O., Girka, V. O., et al. (2008). Propagation of azimuthal waves along the surface of a metal current-carrying cylinder immersed into a magnetized plasma. *Radiophysics and Quantum Electronics, 51*(2), 110–122.

# Chapter 3
# Coupled Surface Flute Waves Propagating in Current-Carrying Plasma Waveguides

To produce and sustain gaseous plasmas in various laboratory devices one can often apply electric currents. Possible flowing of charged plasma particles has not been taken into the account in the Chap. 2, which is devoted to the study of the basic features of transverse eigen modes propagating in simple models of cylindrical waveguides with plasma filling.

In this Chapter the influence of an axial electric current on the dispersion properties of eigen flute modes propagating in various cylindrical waveguide structures with non-uniform plasma filling is under consideration. The plasma filling is assumed here to be sufficiently dense so that the electron cyclotron frequency is lower than the electron plasma frequency. The waveguide is assumed to be uniform along the axial direction. However unlike in the cases studied in the Chap. 2, here the plasma is affected by an azimuthal magnetic field, which is produced by the axial electric current. The value of this axial electric current is supposed to be sufficiently small so that either the azimuthal component of the external magnetic field is much less than its axial component or the cyclotron frequency, which appears due to the azimuthal magnetic field is much less that the eigen frequency of the flute modes in the case if there is no axial magnetic field. But in any case the value of the axial electric current, which is considered here, is sufficiently small to allow one to introduce a small parameter of the problem of coupled flute modes. Thus the tasks considered in this Chapter are studied by the method of successive approximations.

## 3.1 Coupled Low Frequency Surface Flute Waves in Current-Carrying Plasma Waveguide

Let us start our considerations with the case when surface flute waves (SFWs) propagate in a plasma filled waveguide, which is immersed into an external azimuthal magnetic field $\vec{B}_0 = B_{0\varphi}\vec{e}_\varphi$ (there is no axial component of the magnetic

V. Girka et al., *Surface Flute Waves in Plasmas*, Springer Series on Atomic, Optical, and Plasma Physics 79, DOI: 10.1007/978-3-319-02027-3_3,

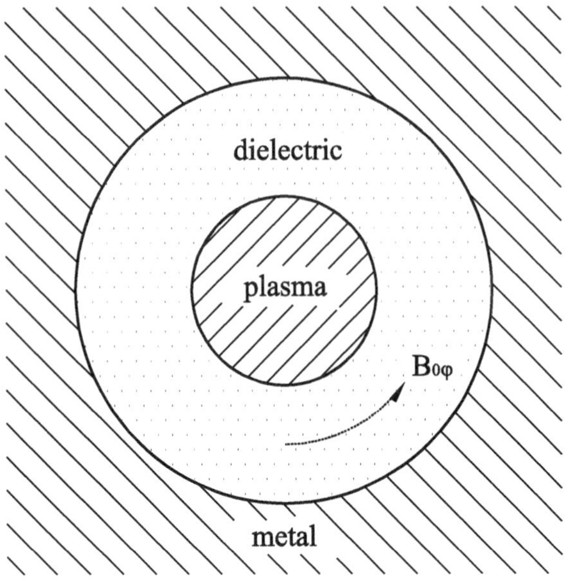


**Fig. 3.1** Scheme of a metal waveguide, which is partially filled with current-carrying plasma in the case of absence of an external axial magnetic field

field: $B_{0z} = 0$), with a frequency that belongs to the LF range (2.3). The geometry of the problem is illustrated in Fig. 3.1.

The value of the azimuthal magnetic field is assumed to be small, so that the inequality $\beta = |\omega_e|/\omega \ll 1$ is valid. In this case the set of the Maxwell equations cannot be represented in the form of two independent sets of equations for two different modes, namely X- and O-modes. Ignoring terms, whose values are smaller than terms of the first order smallness over the small parameter $\beta$, one can derive from Maxwell equations a set of inhomogeneous differential equations for the SFW fields, which consists of two equations, firstly of Eq. (2.19) where the operator, which is located at its right-hand side, has the following form:

$$\widehat{M} = im\varepsilon_2 k/(r\varepsilon_1), \tag{3.1}$$

and secondly, of the equation for the axial component of the SFW magnetic field $H_z^{(m)}$ :

$$\frac{1}{r}\frac{d}{dr}\left(\frac{r}{\varepsilon_1}\frac{dH_z^{(m)}}{dr}\right) + \left(k^2 - \frac{m^2}{r^2\varepsilon_1}\right)H_z^{(m)} = -\widehat{M}E_z^{(m)}. \tag{3.2}$$

The other four components of the SFW field are expressed through the axial components $E_z^{(m)}$ and $H_z^{(m)}$. In the present case ($B_{0z} = 0$), the components of dielectric permeability tensor can be approximately written in the following form: $\varepsilon_1 \approx \varepsilon_3$, $\varepsilon_2 \approx |\omega_e|\,\Omega_e^2\omega^{-3}$ [1]. These equalities are satisfied with an accuracy up to terms of the second order of smallness over the parameter $\beta$. Taking into account expression (3.1), Eqs. (2.19) and (3.2) can be solved using the method of successive approximations.

In this case the boundary conditions for a SFW are determined by:

1. Restriction of the wave fields value inside the considered waveguide, specifically, on the axis of the waveguide;
2. Continuity of tangential components of the SFW fields on the plasma-dielectric interface $r = R_1$;
3. Equality of the SFW tangential electric field value to zero on the metal wall ($r = R_2$) of the waveguide.

Application of these boundary conditions allows one to derive the dispersion relation, which has a typical form for the case of various coupled modes:

$$D_o D_x = D(\beta), \tag{3.3}$$

where $D_o$ is determined by Eq. (2.21) obtained in Sect. 2.4 for the case of ordinary flute modes, $D_x = 0$ is the dispersion relation of the surface flute X-modes (SFXMs) in the case of $B_{0z} = 0$ (see Eq. (2.6) and comments at the end of Sect. 2.1), $D(\beta)$ is the coefficient of coupling between SFXMs and surface flute ordinary polarized modes (SFOMs), whose value is of the second order of smallness over the parameter $\beta$. In the case of non-uniform plasma filling the coefficient of the coupling $D(\beta)$ has the following integral form [1]:

$$D(\beta) = \frac{\varepsilon_1}{\widetilde{\varphi}_o^2 R_1^2 \left(d\varphi_o/dR_1\right)^2} \left[\int_0^{R_1} r\widetilde{\varphi}_o \widehat{M} \varphi_o dr\right]^2, \tag{3.4}$$

here $\varphi_o$ and $\widetilde{\varphi}_o$ are two linearly independent solutions of the homogeneous Eq. (3.2), i.e. for $\widehat{M} = 0$, where $\varphi_o$ has no singularities on the axis of the waveguide unlike $\widetilde{\varphi}_o$.

A solution of the dispersion relation (3.3) can be found in the form of a sum $\omega = \omega_{x,o} + \delta\omega_{x,o}$, where the subscript indexes $x$ and $o$ relate to the cases of these X- and O-modes. So here $\omega_{x,o}$ is the angular SFW frequency calculated in the case of zero axial stationary magnetic field and $\delta\omega_{x,o}$ is the frequency correction caused by the coupling between these two modes. Assuming that the values of the SFW eigen frequencies are known (see Chap. 2), one can find the frequency corrections for SFXMs and/or SFOMs, whose magnitudes are caused by application of a weak external azimuthal magnetic field, in the following form:

$$\delta\omega_{x,o} = \frac{D(\beta)}{D_{o,x}} \left(\frac{\partial D_{x,o}}{\partial\omega}\right)^{-1}_{|\omega=\omega_{x,o}}. \tag{3.5}$$

In most cases the dispersion curves of SFXMs and SFOMs, which propagate in a uniform plasma along the external azimuthal magnetic field, do not intersect with each other. But under specific conditions such crossing is possible. Then the influence of the external field $B_{0\varphi}$ grows up and the ratio $\delta\omega_{x,o}/\omega_{x,o}$ becomes a term of the first order of smallness over the small parameter $\beta$ according to the following formula:

$$(\delta\omega_{x,o})^2 = D(\beta) \left(\frac{\partial D_o}{\partial\omega}\right)^{-1}_{|\omega=\omega_o} \left(\frac{\partial D_x}{\partial\omega}\right)^{-1}_{|\omega=\omega_x}. \tag{3.6}$$

Analyzing the explicit expressions for the frequency corrections, which can be derived from formula (3.5), it is possible to examine the influence of the field $B_{0\varphi}$ on the dispersion properties of SFXMs, which propagate along the surface of an uniform plasma cylinder separated by a narrow vacuum layer $(R_2 - R_1)k \ll 1$ from the metal wall of a cylindrical waveguide. In this limiting case the expression for the eigen frequency of SFXMs is the same as in the case of an isotropic plasma waveguide (one can apply expressions (2.9–2.11), putting there the value of the electron cyclotron frequency $\omega_e = 0$ (see as well the comments at the end of Sect. 2.1). In the limiting case of uniform plasma filling it is found that their frequency correction caused by the field $B_{0\varphi}$ has the following complicated dependence on the radius of the waveguide [1]:

$$\delta\omega_x \approx \beta^2 \omega_x k^2 \phi / (8R_2\Delta), \tag{3.7}$$

where $\delta = c/\Omega_e$ is skin depth and $\phi = \delta^4(R_2 - R_1 + \delta)^{-1}$ if the inequality $R_1 \gg |m|\,\delta$ is valid (limiting case of a wide plasma waveguide). In the opposite limiting case of a narrow waveguide $R_1 \ll \delta$ the factor $\phi$ has the following form: $\phi = R^4(R_2 - R_1 + R_1/|m|)^{-1}(m^2 + |m|^3)^{-2}$. Taking into account the dispersion properties of SFXMs propagating in a plasma filled waveguide under the condition $\vec{B}_{0z} = 0$, one can conclude from the analysis of expression (3.7) that the influence of the external azimuthal magnetic field diminishes with increasing thickness of the vacuum layer $\Delta$ and increasing azimuthal mode number $m$.

In the case of non-uniform plasma density, a region of rarefied plasma can be located nearby the plasma surface. Then the equality $\mathrm{Re}\varepsilon_1(r_0) = 0$ can be satisfied there. Such condition leads to an additional resonant damping of SFXMs. In this case during calculating of the integral, which is represented in expression (3.4) as squared multiplier, one has to take into account the small imaginary part of $\varepsilon_1$ which is caused by collisions between the plasma particles. One can approximately estimate the damping rate $\gamma_r$ of SFXMs under the condition of this resonant damping associated with the existence of the external azimuthal magnetic field:

$$\gamma_r \approx \frac{\pi\beta^2 k^2 \omega_x m^2}{2R_1^2 k_*^5 (R_2 - R_1)(1 + k_*(R_2 - R_1))}, \tag{3.8}$$

where $k_*^3 = k^2 d\varepsilon_1/dR_1 > 0$. Let us underline that SFXM resonant damping, caused by their conversion into ordinary polarized flute modes in the region of rarefied peripheral plasma, where $\mathrm{Re}\varepsilon_1(r_0) = 0$, becomes stronger under the condition of application of an external azimuthal magnetic field $B_{0\varphi}$ than under the condition $B_{0\varphi} = 0$.

Now let us consider the influence of the azimuthal magnetic field on the SFW frequency under the condition of their propagation in a magneto-active plasma waveguide, assuming that the following ratio: $\kappa = B_{0\varphi}/B_{0z} \ll 1$ is the small parameter of the task. For this case the waveguide geometry is shown in Fig. 3.2. Taking into account terms of the first order of smallness over the parameter $\kappa$, one can derive from Maxwell equations a set of differential equations for axial SFW fields $E_z^{(m)}$ and $H_z^{(m)}$. This set consists of the inhomogeneous differential Eq. (2.19)

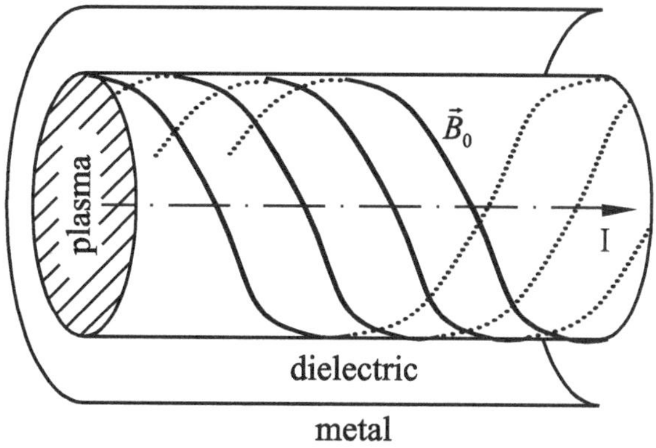


**Fig. 3.2** Scheme of cylindrical metal waveguide, which is partially filled with current-carrying plasma and affected as well by an external axial magnetic field

for the axial component of the SFW electric field, where the operator $\widehat{M}$ now has the following form:

$$\widehat{M} = -ik\kappa\frac{d}{dr} - \frac{k^2\kappa\varepsilon_3}{\varepsilon_1^2 - \varepsilon_2^2}\left(\frac{\varepsilon_1}{ik}\frac{d}{dr} + \frac{m\varepsilon_2}{ikr}\right), \tag{3.9}$$

and of the following equation for their axial magnetic field [1]:

$$\widehat{L}H_z^{(m)} = \widehat{K}E_z^{(m)}, \tag{3.10}$$

where

$$\begin{aligned}\widehat{L} &= \frac{d^2}{dr^2} + \left(\frac{1}{r} + \frac{1}{\varepsilon_1}\frac{d\varepsilon_1}{dr} - \frac{d\ln\left|\varepsilon_1^2 - \varepsilon_2^2\right|}{dr}\right)\frac{d}{dr}\\ &+ \frac{m}{\varepsilon_1 r}\frac{d\varepsilon_2}{dr} - \frac{m\varepsilon_2}{r\varepsilon_1}\frac{d\ln\left|\varepsilon_1^2 - \varepsilon_2^2\right|}{dr} + k^2\frac{\varepsilon_1^2 - \varepsilon_2^2}{\varepsilon_1} - \frac{m^2}{r^2},\end{aligned} \tag{3.11}$$

$$\begin{aligned}\widehat{K} &= \frac{ik}{\varepsilon_1}\left[\frac{-\kappa}{r}\left(\varepsilon_1^2 - \varepsilon_1\varepsilon_3 - \varepsilon_2^2\right) - \kappa\left(\varepsilon_1^2 - \varepsilon_1\varepsilon_3 - \varepsilon_2^2\right)\frac{d}{dr}\right.\\ &- \frac{\kappa m\varepsilon_2\varepsilon_3}{r} - \frac{d\left(\kappa\varepsilon_1^2 - \kappa\varepsilon_1\varepsilon_3 - \kappa\varepsilon_2^2\right)}{dr}\\ &\left.+\kappa\left(\varepsilon_1^2 - \varepsilon_1\varepsilon_2 - \varepsilon_2^2\right)\frac{d\ln\left|\varepsilon_1^2 - \varepsilon_2^2\right|}{dr}\right].\end{aligned} \tag{3.12}$$

One can express the magnetic components $H_r^{(m)}$ and $H_\varphi^{(m)}$ (which are the components of SFOMs in the case $B_{0\varphi} = 0$) through the axial electric component $E_z^{(m)}$ using Eqs. (2.20). The other electric components $E_r^{(m)}$ and $E_\varphi^{(m)}$ of the wave fields are connected with both axial components $E_z^{(m)}$ and $H_z^{(m)}$. With the aid of the method of successive approximation, taking into account the small right hand sides of Eqs. (2.19) and (3.10), one can find their solutions.

After that application of the boundary conditions for the SFW fields on the plasma-vacuum interface allows one to derive the dispersion relation for coupled SFOMs and SFXMs in the form of Eq. (3.3), in which one has to change the expression for $D_x$ by the following one [1]:

$$D_x = \frac{\varepsilon_1 d\psi_o/dR_1 + \varepsilon_2 m\psi_o/R_1}{(\varepsilon_1^2 - \varepsilon_2^2)\psi_o} - k\frac{J_m'(kR_2)N_m'(kR_1) - J_m'(kR_1)N_m'(kR_2)}{J_m'(kR_2)N_m(kR_1) - J_m(kR_1)N_m'(kR_2)}, \tag{3.13}$$

and the coefficient $D(\beta)$ should be replaced there by the following expression as well:

$$D(\kappa) = \frac{-\left(d\varphi_o/dR_1\right)^{-2}}{R_1}\int_0^{R_1} r\varphi_o\widehat{M}\psi_o dr\cdot \left[\frac{\varepsilon_1 W\left(\psi_o, \tilde{\psi}_o\right)}{\left(\varepsilon_1^2 - \varepsilon_2^2\right)\psi_o^2}\int_0^{R_1}\frac{\psi_o\widehat{K}\varphi_o}{W\left(\psi_o, \tilde{\psi}_o\right)}dr + ik\kappa\frac{\varepsilon_1^2 - \varepsilon_1\varepsilon_3 - \varepsilon_2^2}{\varepsilon_1^2 - \varepsilon_2^2}\frac{\varphi_o}{\psi_o}\right], \tag{3.14}$$

where $\psi_o$ and $\tilde{\psi}_o$ are two linearly independent solutions of the homogeneous Eq. (3.10) (i.e. for $\hat{K} = 0$), thus the first of them ($\psi_o$) has no singularities on the axis of the waveguide unlike the second one ($\tilde{\psi}_o$). The Wronskian is $W(\psi_0, \tilde{\psi}_0) = -k_\perp^2/r$ with $k_\perp^2 = k^2(\mu^2 - 1)\varepsilon_1$. All the terms represented in expressions (3.13) and (3.14), which depend on the radial co-ordinate and are not integrants, should be calculated just on the plasma surface $r = R_1$.

One can derive simple analytical expressions for the frequency corrections of SFXMs, which propagate along the interface of a uniform plasma cylinder under the condition of a rather narrow vacuum layer $\Delta \ll 1$ and large plasma density $\Omega_e^2 \gg \omega_e^2$, applying the Eqs. (3.3, 3.5, 3.13 and 3.14). If the plasma cylinder is a narrow one, which means that $\delta_x^2 = k_\perp^{-2} \gg R_1^2$, the expression for the SFXM frequency correction is as follows:

$$\delta\omega_x = \frac{\kappa^2\omega_e^2 R_1^2}{4(|m| + 1)^2 c^2}\left(\Delta + \frac{1}{|m|}\right)^{-1}\left[\frac{m\delta_x^2|\omega_e|}{c^2} + \frac{2\Delta m^2}{\omega_x}\right]^{-1}, \tag{3.15}$$

where the value of the azimuthal component of the magnetic field $B_{0\varphi}$, which determines the magnitude of the dimensionless parameter $\kappa$ should be calculated in the point $r = R_1$. If the radius of the plasma cylinder is sufficiently large (this limiting case can be entitled as "a wide waveguide" because of the validity of the following inequality $R_1 \gg \delta_x$), the SFXM frequency correction is equal to the next expression:

$$\delta\omega_x \approx \frac{\kappa^2\delta_x^4\omega_e^2 m^2}{4R_1^2 c^2(R_2 - R_1 + \delta_0)}\left(1 - \frac{\omega|\omega_e|R_1\delta_x}{mc^2}\right)^2 \left[\frac{m\delta_x^2|\omega_e|}{c^2 R_1} + \frac{2(R_2 - R_1)m^2}{\omega_x R_1^2}\right]^{-1}. \tag{3.16}$$

Carrying out the calculation of (3.15) and (3.16), the field $B_{0\varphi}$ was considered as a linear function of the radial co-ordinate. It should be outlined as well that participation of electrons in axial motion does not lead to a Doppler shift of the frequency of these modes because of the assumption of axial homogeneity of the considered waveguide.

Comparing the expressions (3.15) and (3.16), one can make the following conclusions. First, the influence of the azimuthal magnetic field on the SFXM frequency is stronger in a wide plasma cylinder, than in the case of a narrow plasma cylinder. Second, the frequency correction $\delta\omega_x$ is larger in the case of a plasma-metal interface $R_2 = R_1$, than in the case of utilization of a vacuum gap for separation of the plasma column from the metal wall, $R_2 \neq R_1$. Third, in the case that all other parameters are the same the SFXM eigen frequency is larger under the condition of an applied external azimuthal magnetic field than in the case of its absence ($B_{0\varphi} = 0$) [1].

The eigen frequency of the $n$th radial mode of SFOMs, which propagates in a wide plasma cylinder ($k\Delta \gg |m|\, R_1^{-1}$), can be represented in the form of a sum $\omega_x = \omega^{(0)} + \omega_x^{(1)}$, where the magnitude of the term $\omega_x^{(1)}$ is much smaller than that of $\omega^{(0)}$. In this limiting case, the angular frequency $\omega^{(0)}$ is determined by the characteristics of the vacuum layer and its value is described by the formula (2.23). The frequency correction describes the influence of the characteristics of the magnetized plasma column on the SFW properties and is determined by the following formula:

$$\omega_x^{(1)} \approx \frac{-\omega_x \delta}{R_2 - R_1}\left(1 - \frac{m\delta|\omega_e|}{\omega_x R_1}\right)\left(1 - \frac{m^2}{2k^2R_1^2}\right)^{-1}\left(1 + \frac{m^2}{2k^2R_1R_2}\right)^{-1}_{|\omega=\omega^{(0)}}. \tag{3.17}$$

By equating the frequencies of SFOMs and SFXMs: $\omega_o = \omega_x$, it is possible to find the condition for crossing of their dispersion curves:

$$\frac{R_2 - R_1}{R_1} = \Delta = \frac{\pi\, n\, |\omega_e|}{m\Omega_e}\left(1 - \frac{\omega_e^2 R_1}{2\Omega_e^2 R_2}\right)^{-1}. \tag{3.18}$$

Equation (3.18) can be satisfied only for SFXMs propagating along the direction of the plasma electrons rotation on Larmor orbits nearby the plasma surface. To calculate the frequency corrections of flute electromagnetic perturbations in this case one can apply Eq. (3.6) under the indicated condition (3.18):

$$(\delta\omega)_x^2 \approx \frac{\kappa^2 m^4 c^4 \Omega_e^{-6} \omega_e^4}{16\pi^2 n^2 R_1^2 R_2^2}\left(1 - \frac{m^2}{k^2 R_1 R_2}\right). \tag{3.19}$$

In this case the influence of the field $B_{0\varphi}$ on the SFW frequency enhances with increasing azimuthal mode number $m$, and also with decreasing radial mode number $n$, plasma density and waveguide radius.

From comparison of the results obtained in this section for the cases $B_{0z} = 0$ and $B_{0z} \neq 0$, respectively, one can conclude that the transversal component (in respect to the direction of SFW propagation) of the external magnetic field $B_{0z}$ affects the dispersion properties of SFXMs stronger than the longitudinal $B_{0\varphi}$ component of the applied external magnetic field. Such dependence of the eigen

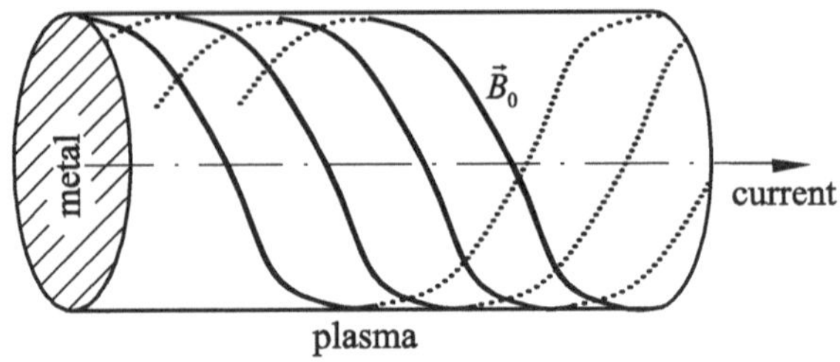


**Fig. 3.3** Model of cylindrical metal antenna surrounded with cold current-carrying plasma in the case $\vec{B}_0 = \vec{B}_{0z} + \vec{B}_{0\varphi}$

frequency on the direction of the external magnetic field coincides with that one, that was obtained in [2] for surface magneto-plasma waves, which propagate along a planar plasma-metal boundary.

## 3.2 Coupled High Frequency Surface Flute Waves Propagating Around a Metal Current-Carrying Cylinder Immersed into Magnetized Plasma

Let us consider an antenna system modeled by a metal cylinder with radius $a$, which is placed in a cold magnetized plasma with a uniform density profile, where the density is sufficiently large so that the inequality $\Omega_e^2 \gg \omega_e^2$ is valid. The external constant magnetic field has two components: $\vec{B}_0 = \vec{B}_{0\varphi} + \vec{B}_{oz}$. The present consideration is restricted to the case of small value of the external azimuthal magnetic field, thus there is a small parameter of the problem: $\kappa = B_{0\varphi}/B_{0z} \ll 1$. The geometry of this case is presented in Fig. 3.3.

Ignoring terms, whose value is higher than the first order of smallness over the small parameter $\kappa$, one can derive the set of differential equations for the axial components of the electric and magnetic fields of SFWs [3] from the Maxwell equations. The equation for the axial electric field looks like (2.19). In the considered case, the operator $\widehat{M}$ located in the right-hand side of this inhomogeneous differential equation is equal to:

$$\widehat{M} = -ik\kappa\frac{d}{dr} - ik\kappa\frac{k_o^2}{k_x^2}\left(\frac{d}{dr} + \frac{m\mu}{r}\right), \tag{3.20}$$

where the following notations are applied: $k = \omega/c$, $k_o^2 = k^2\varepsilon_3$, $k_x^2 = k_\perp^2$ (see as well notations to the Eq. (2.1)). For the axial magnetic field of the coupled SFW one can obtain the following equation:

$$\frac{k_x^2}{r}\frac{\partial}{\partial r}\frac{r}{k_x^2}\frac{\partial H_z}{\partial r} - k_x^2\left[1 + \frac{m^2}{r^2k_x^2} - \frac{m}{r}\frac{\partial}{\partial r}\left(\frac{\mu}{k_x^2}\right)\right]H_z = \widehat{K}E_z, \tag{3.21}$$

where the operator $\widehat{K}$ looks as follows:

$$\widehat{K} = \frac{ik}{\varepsilon_1}\left(\varepsilon_2^2 + \varepsilon_1\varepsilon_3 - \varepsilon_1^2\right)\left[\frac{\kappa}{r} + \kappa\frac{d}{dr} + \frac{\kappa\varepsilon_2\varepsilon_3 m r^{-1}}{\varepsilon_1^2 - \varepsilon_1\varepsilon_3 - \varepsilon_2^2} + \frac{d\kappa}{dr}\right]. \quad (3.22)$$

Let us note that only the wave number $k_o = k\sqrt{\varepsilon_3}$ is present in Eq. (2.19), which describes the flute O-modes in the zero approximation over the small parameter $\kappa$. As far as $k_o^2 < 0$ under the present conditions (because the plasma density is supposed to be sufficiently high), then the fields of these O-modes are determined by cylindrical functions of the first and second kinds [4]. Consequently, the field $E_z$ describes the bulk component of these coupled flute perturbations, and the parameter $k_o$ is their radial wave vector. In zero approximation Eq. (3.21) can be solved in terms of modified cylindrical functions [4]. That is why the radial distribution of the flute perturbations' field $H_z$ has a surface type structure. Therefore the parameter $k_x^{-1} = \left(k\sqrt{(\mu^2 - 1)\varepsilon_1}\right)^{-1}$ determines the penetration depth of the surface component of these coupled flute modes into the plasma. The parameter $k_x$ has a real meaning and its analysis allows one to calculate the limiting bands of the frequency ranges, where SFXMs can exist under the condition when a constant azimuthal magnetic field has zero value, $B_{0\varphi} = 0$.

As it is shown in the Chap. 2, SFXMs with negative azimuthal mode numbers $m$ propagate in the LF range (2.3) and SFXMs with positive azimuthal mode numbers $m$ propagate in the HF range (2.4). An analysis of the $k_o^2$ value proves that in the LF range it is negative, and in the HF range it is positive. That is why the parameter $k_o$ can be referred to be the radial wave number of the bulk component of these coupled flute waves in the HF range. The upper edge of the HF range is determined by the equation $\varepsilon_1 = \varepsilon_2$, that is why nearby this edge the penetration depth of the SFXM into the plasma is considerably larger than the radius of the utilized metal cylinder: $k_x a \ll 1$ (limiting case of a narrow waveguide). The low edge of the HF range is determined by the equation $\varepsilon_1 = 0$, that is why nearby this edge the opposite inequality is realized: $k_x a \gg 1$ (limiting case of a wide waveguide). Results of numerical analysis of the dependence of $k_x^{-1}$ on the value of the metal cylinder radius testify that increasing of the metal cylinder radius leads to decreasing of the penetration depth of these flute waves into the plasma. Reduction of the cylinder radius leads to sharp growing of $k_x^{-1}$, this growth begins sooner for those waves, which have larger azimuthal mode numbers $m$: $k_x^{-1} \approx 2\delta;\ 4\delta;\ 6\delta$ for the $m = 1;\ 2;\ 3$, respectively.

To determine the radial distribution of the SFWs fields in this case one can apply the method of variation of a constant for solving the set of Eqs. (2.19) and (3.21). This allows one to find the expressions for the axial components of the SFW electric and magnetic fields [3]. Application of the boundary conditions indicated in Sect. 2.1 allows one to make the conclusion that SFOMs cannot propagate along the interface between magneto-active plasma and metal. Instead of them, under the condition of application of an external azimuthal magnetic field, the bulk ordinary polarized flute modes (BFOMs) can propagate in such plasma-metal antenna structures. BFOM electromagnetic fields are described by cylindrical functions of the first kind [4], their amplitudes are of the first order smallness over the small parameter $\kappa$ as compared with the SFXM amplitudes.

Using the boundary condition related to the equality of an axial electric field on the interface of the considered plasma-metal antenna system to zero one can derive the following equation:

$$A_1 N_m(k_o a) + A_2 J_m(k_o a) = 0. \tag{3.23}$$

By using absorbing boundary conditions at infinity (see as well Sect. 2.5) one can derive the following relation between the constants of integration $A_{1,2}$ and $B_1$:

$$A_1 + B_1 \frac{\pi}{2} \int_a^\infty r J_m(k_o r) \widehat{M} K_m(k_x r) dr = i[A_2 - B_1 \frac{\pi}{2} \int_a^\infty r N_m(k_o r) \widehat{M} K_m(k_x r) dr]. \tag{3.24}$$

Application of both Eqs. (3.23) and (3.24) makes it possible to express the integration constants $A_{1,2}$ through the constant $B_1$. After that one can apply the boundary condition of equality of the azimuthal component of the electric field of the coupled SFWs to zero on the interface of the metal antenna. This finally leads to the derivation of the following dispersion relation, which describes coupled flute modes propagating along a metal cylinder located in a cold magnetized plasma [3]:

$$D_x^{(0)} + D^{(2)} = 0. \tag{3.25}$$

Here $D_x^{(0)} = 0$ is the dispersion relation of the SFXMs obtained under the condition of zero external azimuthal magnetic field (see Eq. (2.13), where one can put $F_m(x) = K_m(x)$). The summand $D^{(2)}$ is a correction to the SFXM dispersion relation, which is determined by application of an external constant azimuthal magnetic field, its value is of the second order smallness over the small parameter $\kappa$:

$$\begin{aligned} D^{(2)} = & \frac{[k_x K_m'(ak_x) + m\mu K_m(ak_x) a^{-1}]}{J_m(k_o a) + i N_m(k_o a)} \cdot \\ & \int_\infty^a r K_m(k_x r) [N_m(k_o r) \widehat{K} J_m(k_o r) - J_m(k_o a) \widehat{K} N_m(k_o r)] dr \\ & + \frac{\pi}{2} \int_a^\infty r [J_m(k_o r) + i N_m(k_o r)] \widehat{M} K_m(k_x r) dr. \end{aligned} \tag{3.26}$$

It is suitable to search for the solution of Eq. (3.25) in the form: $\omega = \omega^{(0)} + \Delta\omega$, here $\omega^{(0)}$ is the eigen angular frequency of the SFXMs under the condition $B_{0\varphi} = 0$, and $|\Delta\omega| \ll \omega^{(0)}$ is a small correction to their eigen frequency. Then $\Delta\omega$ can be found via solving this dispersion relation, where $D_x^{(0)}$ can be presented as a Taylor series over the parameter $\Delta\omega$. Its imaginary part is negative, since propagation of SFXMs in this case is characterized by losses of electromagnetic energy caused by their conversion into radiated bulk waves. The damping rate $\gamma_{rad} = \text{Im}(\Delta\omega)$ of the SFXM, which is caused by this process is proportional: $\gamma_{rad} \propto D^{(2)}$.

Let us assume in the further consideration within this section that the external azimuthal magnetic field depends on the radial co-ordinate in the following way: $B_{0\varphi}(r) = aB_{0\varphi}(a)/r$. This assumption relates to the case when the axial electric current is flowing only along the surface of the metal antenna, which is immersed into cold plasma, and there is no electric current in the plasma volume.

As it was proved in Sect. 2.2, SFXMs can propagate around a cylindrical metal rod immersed into magneto-active cold plasma in the HF range (2.4) under the condition of absence of an external azimuthal magnetic field, if the inequality $a > m\delta$ is valid. Then the eigen frequency of HF SFXMs can be approximately written in the following form:

$$\omega^{(0)} \approx \sqrt{\Omega_e^2 + \omega_e^2 + m^2c^2 \big/ a^2}. \tag{3.27}$$

In this limiting case of a wide waveguide, the asymptotical expression for the damping rate of these coupled flute waves caused by generation of electromagnetic radiation can be represented in the following form:

$$\gamma_{rad} \approx \kappa(a)^2 m^3 \delta^3 a^{-3} \omega^{(0)}. \tag{3.28}$$

It is interesting to compare this case with that one considered in the previous section. The frequency of SFXMs propagating in plasma filled waveguides in the HF range under the condition of absence of an external azimuthal magnetic field is characterized by both imaginary part (see expression (3.28)) and real part $\text{Re}(\Delta\omega)$ of the frequency correction. Under this condition $\text{Re}(\Delta\omega)$ for coupled flute waves can be calculated from the expression (3.5):

$$\text{Re}(\Delta\omega) = a \cdot \gamma_{rad}/\delta. \tag{3.29}$$

As it was found in the Chap. 2 the collisional damping rate of these surface waves is proportional to the effective value of the collision frequency $\nu$ (see expression (2.27)). Comparing this value with (3.28), it is possible to determine the conditions, when the collisional damping rate is larger than the damping rate caused by radiation of the coupled flute waves' electromagnetic energy:

$$\nu/\omega^{(0)} \gg \kappa^2 (m\delta/a)^3. \tag{3.30}$$

Let us estimate the conditions, under which the collisional damping rate will be less than the damping rate caused by radiation of the coupled flute modes electromagnetic energy, on the base of inequality (3.30) for the case of an antenna with a radius, which is twice as large as the product $m\delta$. In the case, when $\nu$ is five orders less than the eigen frequency of the coupled flute modes, for $\kappa = 0.01$ the collisional damping rate will be less than the damping rate caused by SFW power radiation due to bulk O-mode emission. Remind that if the plasma density is $n_e \approx 10^{10}$ cm$^{-3}$, the frequency of collisions between cold plasma particles is about $\nu \approx 10^4$ s$^{-1}$ [5, 6].

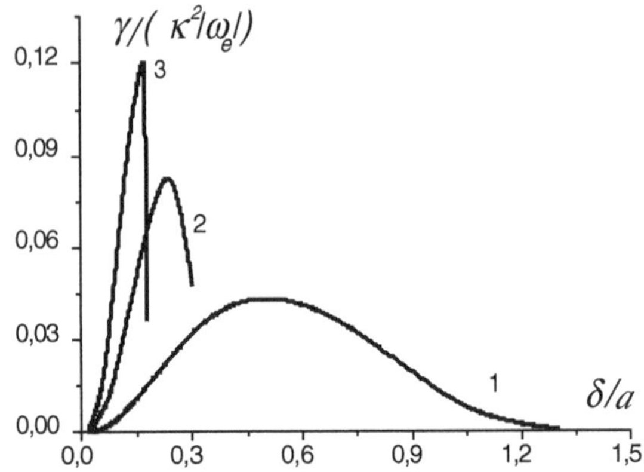


**Fig. 3.4** Normalized damping rates of coupled flute modes versus normalized reversed radius of the metal antenna immersed into plasma; the numbers indicate the values of the azimuthal mode number *m*

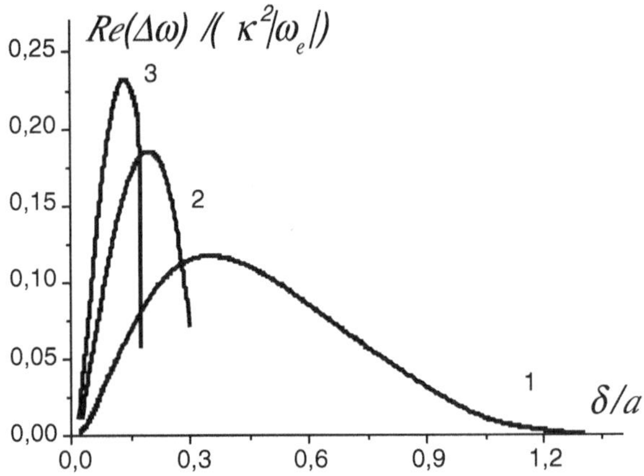


**Fig. 3.5** Real parts of the correction to the angular frequency of coupled flute modes propagating around the metal antenna immersed into plasma as function of reverse radius of this antenna; the numbers indicate the values of the azimuthal mode numbers *m*

The results of the numerical analysis of dispersion relation (3.25) allow one to make the following conclusions:

1. The dispersion curves of SFXMs with different azimuthal mode numbers only weakly differ from each other in the case of metal antennas with large radius;
2. Increasing of the azimuthal mode number leads to narrowing of the range of waveguide radius, where SFXM propagation is possible.

The following four diagrams show the results of the analysis. In Fig. 3.4 the results of numerical calculation on the value of the damping rate of the coupled flute modes caused by conversion of SFXMs into BFOMs are presented. It is shown that decreasing azimuthal mode number $m$ leads to decreased damping of the coupled flute X- and O-modes, this feature is connected with the presence of an azimuthal component of the external magnetic field. The maximal value of this damping rate grows linearly with increasing value of $m$.

The results shown in Fig. 3.5 prove that the larger the value of $m$ is, the stronger is the influence caused by the application of the field $B_{0\varphi}$ on the dispersion properties of coupled flute modes. The maximal absolute value of the real part of the frequency correction $\Delta\omega$ is larger than the maximal absolute value of the corresponding damping rate $\gamma_{rad}$.

The numerical analysis shows that the radial distribution of the real part of the correction to the azimuthal magnetic field of the coupled flute modes, which is determined by application of the field $B_{0\varphi}$ is approximately similar to the graph of the first kind Bessel function $J_n(x)$ as well [4]. Its absolute value is proportional to the parameter $\kappa$, so this correction is of the first order of smallness. The axial electric field of the coupled flute waves is also approximately similar to the graph of the function $J_n(x)$.

The results of the numerical analysis of the radial distribution of the coupled flute modes fields are depicted in Figs. 3.6 and 3.7. The radial distribution of the azimuthal magnetic field of the coupled SFWs in zero approximation is shown by the bold line in Fig. 3.6. The dashed line indicates the radial distribution of the real part of the first order (over small parameter $\kappa$) correction to the azimuthal magnetic field of these waves, which is determined by application of the field $B_{0\varphi}$. As one can see from Fig. 3.7, the radial wavelength of the BFOMs $\lambda \approx 6.13a$ relates to the meaning of the product $k_o\lambda \approx 2\pi$; the absolute value of the axial electric field of the coupled SFW is approximately of the same order as the correction to the waves' azimuthal magnetic field $\Delta H_z$ near-by the antenna up to a distances of $r \approx 10a$.

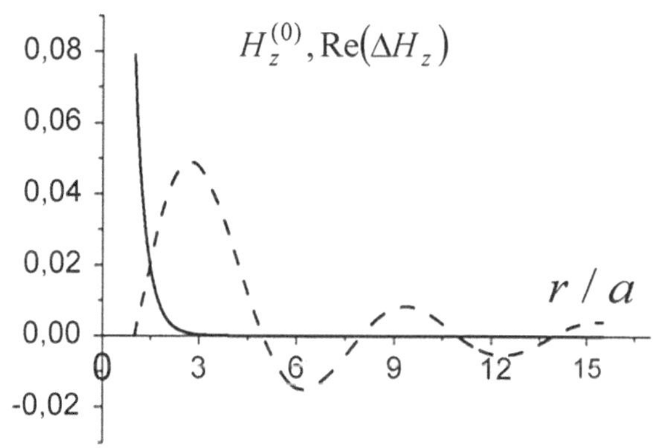


**Fig. 3.6** Radial distribution of $H_z^{(0)}$ and $\mathrm{Re}(\Delta H_z)$ of coupled flute modes excited by a metal antenna immersed into plasma for $\kappa(a) = 1$

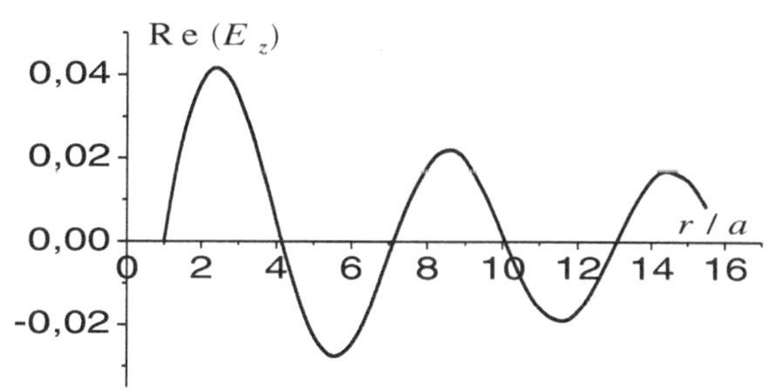


**Fig. 3.7** Radial distribution of $\mathrm{Re}(E_z)$ of coupled flute modes excited by a metal antenna immersed into plasma

To make a conclusion on the possibility to observe the coupled flute X- and O-modes in practical experiments, let us perform the following estimations. For instance, if a gaseous plasma is sustained in the experimental device under the following conditions: $n_e \approx 10^{10}\,\mathrm{cm}^{-3}$ and $B_{0z} \approx 0.0064\,T$, then for the coupled flute modes with $m = 1$ one can obtain $\Omega_e = 5\,|\omega_e|$, which is a little bit less than the eigen angular frequency $\omega^{(0)} \approx 6 \cdot 10^9\,\mathrm{s}^{-1}$. Under the condition of application of an external axial current $I \approx 24\,A$, which produces an azimuthal magnetic field $B_{0\varphi} \approx 45 \cdot 10^{-6}\,T$, the small parameter of the problem is $\kappa \approx 7 \cdot 10^{-3}$, and one can find the frequency correction $\mathrm{Re}(\Delta\omega) \approx 5700\,\mathrm{s}^{-1}$. In this case the damping rate determined by emission of the BFOM is $\gamma_{rad} \approx 2400\,\mathrm{s}^{-1}$. This means that the value of $\gamma_{rad}$ is maximum for $a \approx 2\delta \approx 10.6\,cm$. To estimate the frequency of ion–electron collisions we have applied the formulas obtained in [6]: $\nu \approx 25 n_e T_e^{-3/2}$ (here the density of plasma particles is calculated in $cm^{-3}$ and the electron temperature $T_e$ in $K$), so for the temperature $T_e \approx 5$ eV one can obtain $\gamma_{col} \approx 9000\,\mathrm{s}^{-1}$, which is larger than both values of $\gamma_{rad}$ and frequency correction $\mathrm{Re}(\Delta\omega)$. The Q-factor of such oscillations is sufficiently large: $Q = \omega^{(0)}/(2\gamma_{col}) \approx 7 \cdot 10^5$.

But increasing the plasma temperature up to value $T_e \approx 50$ eV leads to decreasing value of the collisional frequency. It becomes approximately thirty times less as compared with the value indicated above. Then the damping rate $\gamma_{rad}$ is about $10\gamma_{col}$ and the value of the Q-factor also increases by a factor of 3.5. Therefore experimental investigations into the coupling between SFXMs and BFOMs can be carried out without utilization of a somewhat complicated apparatus or extreme experimental conditions.

## 3.3 Coupled High Frequency Surface Flute Waves Propagating in Waveguide Entirely Filled with Current-Carrying Plasma

Let us consider a cylindrical, ideally conducting metal waveguide with radius $a$, which is entirely filled with cold plasma. Its density is assumed to be uniform along axial and azimuthal directions, but its radial profile could be non-uniform, in

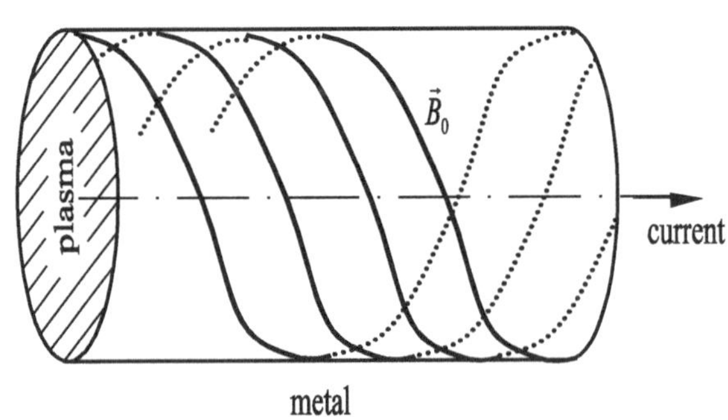


**Fig. 3.8** Scheme of cylindrical metal waveguide, which is entirely filled with non-uniform current-carrying plasma

the general case. Thus the dispersion equation for coupled flute waves is obtained for the general case in this section, but analytical solutions have been obtained for the limiting case of plasma with uniform density. The waveguide is assumed to be immersed into an external steady magnetic field $\vec{B}_0 = \vec{B}_{0z} + \vec{B}_{0\varphi}$. As it was supposed in the previous section, here the case of small external azimuthal magnetic field is under consideration, so that the inequality $\kappa = B_{0\varphi}/B_{0z} \ll 1$ is valid. The absence of a dielectric layer between the metal wall of the waveguide and the plasma column makes propagation of SFOMs impossible and ordinary polarized flute modes propagate as bulk modes. The geometry of this case is presented in Fig. 3.8.

To find the radial distribution of the flute waves' electromagnetic fields the set of Maxwell equations is applied. The presence of an azimuthal component $B_{0\varphi}(r)$ of the external constant magnetic field leads to the situation, that in cylindrical co-ordinates $(\vec{e}_r, \vec{e}_\varphi, \vec{e}_z)$ just all nine elements of the dielectric permeability tensor $\varepsilon_{ik}$ of a cold magneto-active plasma are non-zero. That is why solving the Maxwell equations becomes in this case a rather complicated problem as compared with the cases that are studied in Chap. 2.

To simplify the task one can choose a special local co-ordinate system. Let us apply here an ortho-normal basis of the co-ordinate vectors $(\vec{e}_1, \vec{e}_2, \vec{e}_3)$, which are connected with the force lines of the external magnetic field $\vec{B}_0$. The first vector of this ortho-normal basis coincides with the radial basis vector in cylindrical co-ordinates, $\vec{e}_1 = \vec{e}_r$; the third vector is parallel to the magnetic force lines, $\vec{e}_3 = \vec{B}_0/\left|\vec{B}_0\right|$; and the second vector adds these three vectors to a right-handed coordinate system: so the following expression $\vec{e}_2 = \vec{e}_3 \times \vec{e}_1$ is valid. Within the accuracy up to terms of the second order of smallness over the small parameter $\kappa \ll 1$, which characterizes the problem, one can obtain the following approximate relations between the proposed new co-ordinate vectors and the basis co-ordinate vectors of a cylindrical system: $\vec{e}_1 = \vec{e}_r$, $\vec{e}_2 \approx \vec{e}_\varphi - \kappa\,\vec{e}_z$, $\vec{e}_3 \approx \vec{e}_z + \kappa\vec{e}_\varphi$. Using these local co-ordinates, one can write down the set of Maxwell equations, which incorporates the following simple relation between the components of the displacement vector $\vec{D}$ and the electric field $\vec{E}$ of the wave:

$$\vec{D} = \varepsilon_1 \left(E_1\vec{e}_1 + E_2\vec{e}_2\right) + \varepsilon_3 E_3\vec{e}_3 - i\varepsilon_2\vec{e}_3 \times \vec{E}. \tag{3.31}$$

In this case the dielectric permeability tensor of a cold magneto-active plasma with five non-zero components [7–9] can be applied. The value of the cyclotron frequency in the expressions for these components of the tensor $\varepsilon_{ik}$ is determined by the complete constant magnetic field $\vec{B}_0$, whose value with accuracy up to small terms of the second order of smallness over $\kappa$ is equal to the axial magnetic field $B_0 \approx B_{0z}$. In the applied right-handed coordinate system $(\vec{e}_1, \vec{e}_2, \vec{e}_3)$, components of the rotor-operator within accuracy up to terms of the second order of smallness $\sim \kappa^2$ can be expressed through the components of this operator calculated in cylindrical co-ordinates in such way:

$$rot_1 = rot_r,\ rot_2 \approx rot_\varphi - \kappa\,rot_z,\ rot_3 \approx \kappa\,rot_\varphi - rot_z. \tag{3.32}$$

Taking into account all observations mentioned above, one can obtain the following three scalar equations from the set of Maxwell equations, which in zero approximation describe the ordinary polarized electromagnetic waves:

$$-mH_z/kr = \varepsilon_1 E_r + i\varepsilon_2(E_\varphi - \kappa E_z), \tag{3.33}$$

$$\frac{idH_z}{kdr} + \frac{i\kappa}{kr}\frac{d\left(rH_\varphi\right)}{dr} + \frac{\kappa m}{kr}H_r = i\varepsilon_2 E_r - \varepsilon_1 E_\varphi + \kappa\varepsilon_1 E_z, \tag{3.34}$$

$$-\frac{i\kappa}{k}\frac{dH_z}{dr} + \frac{i}{kr}\frac{d(rH_\varphi)}{dr} + \frac{m}{kr}H_r = \varepsilon_3(E_z + \kappa E_\varphi). \tag{3.35}$$

Solving the set of Eqs. (3.33–3.35) for the axial electric field, one can derive an inhomogeneous differential equation of the second order, which has the form of Eq. (2.19) with small right-hand part. It looks like the expression (3.20). The other three equations of the set of Maxwell equations in zero approximation describe extraordinary polarized surface waves. They also can be reduced to an inhomogeneous differential equation of the second order like Eq. (3.21), where the influence of the constant azimuthal magnetic field shows up by the appearance of a small right-hand term like expression (3.22). The radial wave number of the BFOMs and the penetration depth of the SFXMs are determined in the previous Sect. 3.2 (see Eq. (3.20) and the notations for $k_o^{-1}$ and $k_x^{-1}$ there).

As far as these waves are described by equations which are typical for weakly coupled modes it is suitable to apply the method of successive approximations, using the theory of SFXMs as the zero approximation for their study. Solutions of Eqs. (2.19) and (3.21) with their right-hand sides indicated above can be found by the method of constants variation. The fields of the investigated waves should satisfy the boundary conditions, which are indicated in Chap. 2. Thus their application leads to the derivation of a dispersion relation for the coupled flute modes likes Eq. (3.3). But in the present case the factor $D_o$ which is written there is determined by formula (2.21), where one should take $R_1 = R_2 = a$. Using factor the $D_x$ from the Eq. (3.3) one can derive the dispersion equation of the SFXMs in absence of an external constant azimuthal magnetic field $D_x = 0$. In the present case, an explicit expression for $D_x$ can be written as follows [10]:

$$D_x = -\frac{k^2}{k_x^2}\left(\frac{1}{\psi_x}\frac{d\psi_x}{dr} + \frac{m\mu}{a}\right). \tag{3.36}$$

Under the present conditions the summand $D(\kappa)$ in Eq. (3.3) is the coupling coefficient which describes the interaction between BFOMs and SFXMs. Its value is quadratic over the small parameter $\kappa$:

$$D(\kappa) = \left(\frac{d\varphi_o}{dr}\right)^{-2}\int_0^a \frac{r}{a}\varphi_o\widehat{M}\psi_x dr\left[\frac{k^2 W\left(\psi_x, \tilde{\psi}_x\right)}{k_x^2\psi_x^2}\int_0^a \frac{\psi_x\widehat{K}\varphi_o dr}{W\left(\psi_x, \tilde{\psi}_x\right)} - i\kappa k\left(1 + \frac{k_o^2}{k_x^2}\right)\frac{\varphi_o}{\psi_x}\right]. \tag{3.37}$$

The functions $\varphi_o$ and $\psi_x$, which are utilized in formulas (3.36) and (3.37), are the solutions of Eqs. (2.19) and (3.21) with zero right-hand sides. Both functions should be of limited values within all the volume of the waveguide, in particular, in the centre of the plasma column ($r = 0$). The parameters, which depend on the radial co-ordinate but are not integrand in the expression (3.37), should be calculated at $r = a$.

Solutions of the dispersion relation, which describe coupled flute X- and O-modes propagating in the waveguide entirely filled with cold magneto-active plasma can be found in the form $\omega = \omega_{O,X} + \delta\omega_{O,X}$, where $\omega_{O,X}$ are the eigen frequencies of BFOMs and SFXMs, respectively, calculated in the case of absence of an external azimuthal magnetic field, $|\delta\omega_{O,X}| \ll \omega_{O,X}$. The corrections $\delta\omega_{O,X}$ to the eigen frequencies are determined by formula (3.5) in this case.

Simple analytical solutions of the dispersion relation for coupled BFOMs and SFXMs can be obtained in the limiting case of uniform dense plasmas. In this case Eqs. (2.19) and (3.21) for the electromagnetic fields $E_z$ and $H_z$ are non-uniform Bessel equations. Solutions of the corresponding uniform equations, which are limited in the centre of the plasma column, can be expressed in terms of Bessel functions of the first kind [4] with the argument $k_o r$ and modified Bessel functions [4] with the argument $k_X r$, respectively. An analysis of the expression for the parameter $k_X$ allows one to determine frequency ranges, where SFXM can exist. The analysis of $k_o^2$, which determines features of the flute O-modes testifies that it has different signs in the HF and the LF ranges, namely, in the LF range (2.3) the inequality $k_o^2 < 0$ is valid, and in the HF range (2.4) it becomes positive: $k_o^2 > 0$. Thus, linear interaction between BFOMs and SFXMs is possible only in the HF range (2.4) for waveguides, which are entirely filled with cold magneto-active plasma [10].

Let us consider the case of linear dependence of the external constant azimuthal magnetic field on the radial co-ordinate: $B_{0\varphi}(r) = rB_{0\varphi}(a)/a$. Then the integrals, which are included in the expression (3.37) for the parameter of X- and O-modes coupling $D(\kappa)$, can be calculated in an explicit form [10]. An analysis of the SFXM dispersion relation proves that in narrow waveguides HF SFXMs cannot propagate. In wide waveguides ($a \gg \delta\,|m|$) the eigen frequency $\omega_X$ of HF SFXMs is approximately determined by the expression (3.27) given in the previous section.

If the values of the applied magnetic field and/or the waveguide radius are sufficiently small, so that the following inequality is valid:

$$a^2\omega_e^2 \big/ c^2 < g_{|m|,s}^2 - m^2, \tag{3.38}$$

BFOMs cannot propagate in the HF range (2.4). Here $g_{|m|,S}$ is the S-th root of the first kind Bessel function of the order $|m|$ [4]. With increasing radius of the utilized waveguide and/or increasing value of the external constant magnetic field, inequality (3.38) becomes invalid. Then one can obtain a corresponding equality instead of inequality (3.38). The validity of this equality will mean that the frequencies of SFXMs and BFOMs are equal to each other under the corresponding condition.

Therefore in this case, the correction to the SFXM eigen frequency will be of the following form:

$$\delta\omega_X \approx \omega c^2 a^{-2}\sqrt{m^3\omega_e\Omega_e^{-3}} \approx \kappa\,|m|^{3/2} Z^{-5/2}\Omega_e/g^2_{|m|,s}, \tag{3.39}$$

here $Z = \Omega_e/|\omega_e| > 1$. From the analysis of expression (3.39), one can conclude that the influence of $B_{0\varphi}$ on the frequency of SFXMs increases with increasing axial constant magnetic field, and also with decreasing plasma density, absolute value of azimuthal mode number $m$ and radius of the waveguide [11].

To investigate the dispersion properties of the coupled flute waves for arbitrary values of the parameters of the waveguide system one can apply a numerical method. It allows one to calculate, for instance the crossing-point of SFXMs with azimuthal mode number $m = -1$ and the first three radial BFOM modes. In Fig. 3.9 the dispersion curve of SFXMs (depicted by bold line) and also the dispersion curves, which illustrate the first, second and third radial modes of BFOMs (dashed lines marked by numbers 1, 2 and 3, respectively), which propagate with azimuthal mode number $m = -1$ under the condition of zero field $B_{0\varphi}$, are presented as functions of the effective wave number $k_{ef} = |m|\,c/(a\Omega_e) = |m|\,\delta/a$. For the following values of the waveguide parameters: $\kappa = 0.1, \mathrm{Z} = 3$ the crossing-points can be approximately found from expression (3.38): $k_{ef} \approx |m|\,/(Z\,g_{|m|,s}) \approx 0.087, 0.048, 0.033\ldots$. To see the fine structure of the frequency spectrum, which illustrates interaction between SFXMs and BFOMs one can apply the dimensionless variable $y = \sqrt[4]{\omega^2\omega_e^{-2} - 1 - Z^2}$, where $y = 0$ indicates the lower edge of the HF range (2.4).

Let us consider the variation of the fine structure of the spectral interaction between the SFXMs and the BFOMs with decreasing effective azimuthal wave number $k_{ef}$. One can see from Fig. 3.10 that the dispersion curve for the SFXM (curve *1*) comes from the range of maximal effective wave numbers and the function $y(k_{ef})$ undergoes a discontinuity near $k_{ef} \approx 0.21$. On the left side of this discontinuity, the dispersion curve for the SFXM emerges from the low boundary of the frequency range (Fig. 3.10, right-hand branch of curve *2*). Near the point $k_{ef} \approx 0.087$, the SFXM transforms into the BFOM and this dispersion curve then

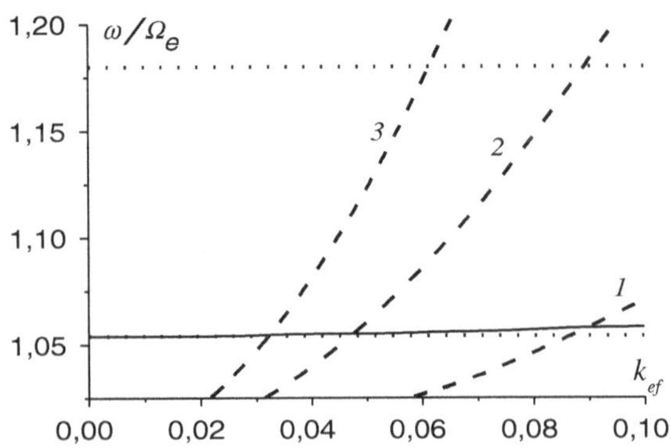


**Fig. 3.9** Dependence of the normalized angular frequency of coupled flute modes on the effective wave number for the case of a cylindrical waveguide entirely filled with plasma. The dotted lines indicate the limits of the HF range

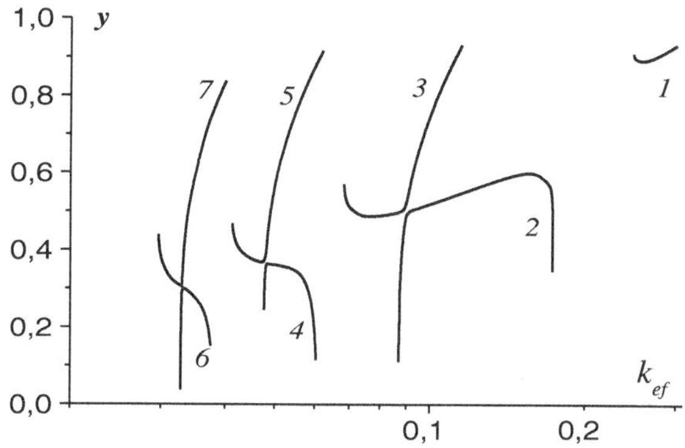


**Fig. 3.10** Fine structure of interaction between SFXMs and three radial modes of BFOMs propagating in cylindrical waveguide entirely filled with plasma. The branches of the curves marked by the numbers 1 ÷ 7 are explained in the text (see Sect. 3.3)

abruptly falls to the boundary of the frequency range (left-hand branch of curve 2). The dispersion curve, which is drawn for the first radial bulk flute O-mode (right-hand branch of curve 3) comes from the range of relatively high effective wave numbers. Near the point $k_{ef} \approx 0.087$, the BFOM transforms into the SFXM and then this curve runs almost horizontally to the left and rises rapidly when it approaches to $k_{ef} \approx 0.064$ (left-hand branch of curve 3) and so on in the same manner. One can see in Fig. 3.10 the similarity in behavior of the curves marked by 3, 5 and 7; the same can be said on the behavior of the curves marked by numbers 2, 4 and 6, respectively.

Therefore, one can make the following conclusions concerning the interaction between the surface flute X-modes and bulk flute O-modes, if the radial wave number of the BFOM grows up:

1. The difference between the dispersion curves of BFOMs and SFXMs becomes less and less pronounced;
2. The range of the $k_{ef}$ values between two neighboring modes conversion crossing points of the dispersion curves for SFWs becomes narrower;
3. Reduction of the $k_{ef}$ value leads to a decrease of the SFXM frequency towards the lower edge of the HF range.

Thus interaction between SFXMs and BFOMs results in the appearance of forbidden bands in the flute waves frequency spectrum, which are typical for waves propagating in media with periodic spatial non-uniformity (see, for example [12]). The appearance of the forbidden bands in the frequency spectrum, which is shown in Fig. 3.10, is equivalent to the existence of ranges of forbidden $k_{ef}$ values. This is explained by the fact that for these values of the plasma cylinder radius, the BFOMs stop to be the eigen modes and approaching to these forbidden ranges of $k_{ef}$ leads to a sharp increasing of the correction term to the frequency of BFOMs, which does not allow further application of the method of successive approximation.

One also can see the appearance of $k_{ef}$ ranges, within which SFXMs are characterized by reverse dispersion ($d\omega/dk_{ef} < 0$), which is caused by the interaction between SFXMs and BFOMs. This means that in waveguides with the corresponding values of parameters, flute wave energy is transported in the direction, which is opposite to the direction of the SFW phase velocity (backward waves).

It should be underlined that the points $k_{ef} \approx 0.260, 0.160, 0.075, \ldots$, on the dispersion curve shown in Fig. 3.10, where direct (or in other word, normal) dispersion changes to reverse (abnormal) dispersion (or vice versa) and where $d\omega/dk_{ef} = 0$, are of particular interest. Nearby these points the dependence of the frequency of flute waves $\omega = \omega(k_{ef})$ can be represented in the following form:

$$\omega \approx \omega_0(1 + \alpha^2 k_{ef}^2/2). \tag{3.40}$$

So if at the initial time, the starting value of an electromagnetic pulse of the studied SFW has the form of a modulated sinusoidal function with Gaussian envelope:

$$H_z(\varphi, 0) = H^{(0)} \cos(m\varphi) \exp[-\varphi^2/(2\beta^2)], \tag{3.41}$$

then at any next moment of time this electromagnetic pulse will be of a modulated sine-wave form, but the effective width $\beta_t$ of its envelope, which is determined by a Gaussian curve, will depend on time in the following way:

$$\beta_t^2 = \beta^2 + \omega_0 \alpha^2 \delta^2 t/a^2. \tag{3.42}$$

Thus the form of the electromagnetic pulse coincides with its initial form only during the following specific short interval of time: $t \ll a^2\beta^2\alpha^{-2}\delta^{-2}$ [13].

Let us estimate the possibility of experimental realization of the phenomenon described above. For a laboratory gaseous plasma with the density $n_e \approx 10^{11}\,\mathrm{cm}^{-3}$ it is necessary to apply the axial magnetic field $B_{0z} \approx 0.048\,T$ (then Z = 3). Coming from condition (3.38), one can find that the minimum value of the plasma radius, which is necessary for achieving the resonant interaction between SFXMs and BFOMs with azimuthal mode number $m = -1$, will be equal to $a \approx 12$ cm. To satisfy the condition $B_{0\varphi} = 0.1\,B_{0z}$ one has to apply the axial electric current $I \approx 3$ kA. Then the difference between the angular frequencies nearby the crossing point of the dispersion curves of a SFXM and the first radial mode of BFOMs will be equal to $1.4 \times 10^7\,\mathrm{s}^{-1}$.

Summarizing the results of numerical analysis of coupled flute modes, one can see that if the condition of linear interaction between BFOMs and SFXMs is not satisfied, the frequency correction to the SFXM eigen frequency is proportional to the external azimuthal magnetic field value raised to the second power, while the corresponding correction determined by a constant axial magnetic field $B_{0z}$ is proportional to its value (i.e. $\delta\omega_x \propto B_{0z}$). Linear resonant interaction between SFXMs and BFOMs is realized nearby the crossing points of their dispersion curves. It is accompanied by the appearance of ranges of forbidden bands of their frequencies. In this case, the frequency correction becomes larger: it is directly proportional to the external azimuthal magnetic field.

It was found, however, that a constant azimuthal magnetic field $B_{0\varphi}$ has the greatest effect on the dispersion properties of SFXMs by no means at the points of intersection between the dispersion curves of these surface X- and bulk O-modes.

It should be noted that the waveguide radius accommodates an almost integer number of BFOM half-wavelengths near these points on the presented plot. Therefore, as in the case of linear SFXM–BFOM interaction, the boundary conditions are satisfied owing to a small variation of the frequency, because the field $B_{0\varphi}$ has a strong effect on the radial distributions of both modes. Since the BFOM field distribution is of bulk type, one can select sets of waveguide parameters such that in zero approximation the plasma waveguide radius can accommodate an odd number of quarters of the wavelength of the induced BFOM. Such conditions are observed almost in the middle between the points where the dispersion curves for the SFXM intersect with corresponding radial BFOM harmonics. In this situation, the plot of the dispersion curves for the coupled SFXMs and BFOMs is characterized by the appearance of forbidden bands of the wave numbers. Then, to satisfy the boundary conditions, the frequency correction must be so high that the corresponding increase (or decrease) in the wavelength removes (adds) an excessive (missing) quarter-wavelength. This also results in a considerable change in the SFXM dispersion in wave number ranges adjacent to the forbidden $k_{ef}$ bands. In these ranges, the SFXM dispersion becomes of reverse type; i.e., the energy of these flute waves is transmitted in the direction opposite to their phase velocity vector. It illustrates the qualitative difference between SFXM–BFOM linear interaction and interaction of SFXMs with the surface flute ordinarily polarized modes.

Let us compare the present results with those obtained in the previous Sect. 3.2. The values of SFXM frequencies, which propagate in metal waveguides entirely filled with plasma, are very close to the frequencies of SFXMs, which propagate along the boundary between a cold magneto-active plasma and a metal rod immersed into it (see Sect. 2.2). But the influence of a weak external constant azimuthal magnetic field on their dispersion properties appears to be different in these two cases. In the case of a metal waveguide, which is entirely filled with plasma [10], resonant linear interaction between SFXMs and BFOMs can be realized. This phenomenon is not observed in the case of the coupled flute waves, which propagate along the boundary of a metal rod immersed into plasma (just this case was studied in Sect. 3.2). On the other hand, linear interaction between SFXMs and BFOMs in a metal waveguide that is entirely filled with a cold magneto-active plasma is not accompanied by damping of the coupled flute modes. While their interaction under the condition of their propagation along the boundary of the metal rod immersed into a cold magneto-active plasma results in flute waves damping that is caused by conversion of SFXMs into BFOMs, which are radiated out of this antenna—plasma system. Thus the amplitude of the waves, which are radiated from the plasma coated antenna, is linear over the strength of the electric current flowing along the antenna and the damping rate of the flute waves is quadratic over the strength of electric current.

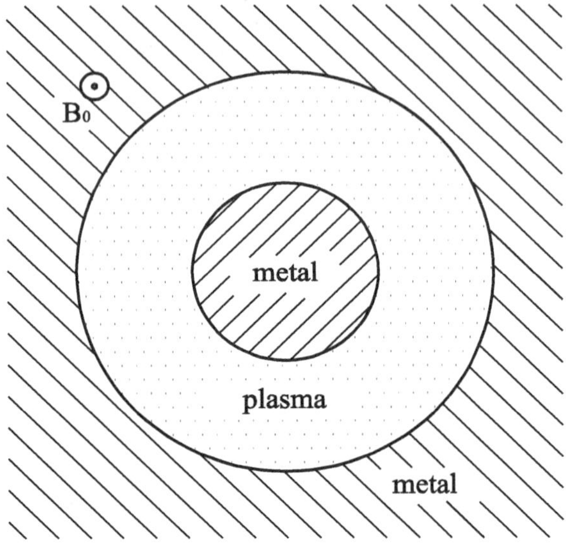


**Fig. 3.11** Scheme of coaxial metal waveguide, which is entirely filled with magneto-active current-carrying plasma

## 3.4 Coupled High Frequency Surface Flute Waves in Coaxial Plasma Waveguide Entirely Filled with Current-Carrying Plasma

To increase the efficiency of operation of high-frequency electronic devices just coaxial waveguide structures are utilized more and more frequently because their characteristics are found to be better than the best ones of cylindrical structures. Therefore, let us examine theoretically a coaxial metal waveguide with inner and outer radii $a$ and $b$, respectively. The space between the metal walls of the coaxial waveguide is entirely filled with magneto-active plasma that has a uniform density profile. The geometry of this coaxial waveguide is presented in Fig. 3.11. Let us consider the case when $\kappa = B_{0\varphi}/B_{0z} \ll 1$. This allows one to carry out not only numerical but also analytical studies into the properties of coupled flute waves, which can propagate in such waveguides [14].

Ignoring the summands, whose order of smallness is higher than $\kappa$ value, one can derive the set of differential equations for the axial components of the electric field $E_z$ of BFOMs and magnetic field $H_z$ of SFXMs from the Maxwell equations. The tangential components of the electric field of the considered electromagnetic perturbations are assumed to be equal to zero on the metal surfaces of the coaxial line. The frequency ranges for the existence of these flute waves and the signs of their azimuthal mode numbers are determined in Chap. 2. To study the dispersion properties of such coupled flute electromagnetic waves let us make use of the method of successive approximations, applying the theory of SFXMs (see Sect. 2.2) as the zero approximation. In this case one can find solutions of the Maxwell equations for the axial electric and magnetic fields of flute electromagnetic oscillations by the method of variation of constants [14]. All the other components of the electromagnetic field of the coupled flute modes can be expressed through the $E_z$ and the $H_z$ components from the Maxwell equations.

Application of the boundary conditions indicated in Chap. 2 allows one to derive the dispersion relation in the form (3.3), where $D_o = J_m(k_o a)N_m(k_o b) - N_m(k_o r)J_m(k_o r) = 0$ is the dispersion relation of BFOMs and $D_x = 0$ is the dispersion relation of SFXMs (both these equations are obtained for zero azimuthal external magnetic field, $B_{0\varphi} = 0$). In the present case, the SFXM dispersion relation looks like follows:

$$D_x = a_{23}a_{44} - a_{24}a_{43}. \tag{3.43}$$

In the dispersion equation obtained in [14] for the case of coaxial plasma filled waveguide, the parameter $D(\kappa)$ plays the role of the coupling coefficient between BFOMs and SFXMs that describes the influence of the azimuthal component of an external magnetic field on the propagation of these modes:

$$\begin{aligned} D(\kappa) = \frac{\pi}{2} \Bigg\{ & \left[ a_{23} \int_a^b r I_m(k_x r) \widehat{K} N_m(k_o r)\, dr - a_{24} \int_a^b r K_m(k_x r) \widehat{K} N_m(k_o r)\, dr \right] J_m(k_o r) \\ & - N_m(k_o r) \left[ a_{23} \int_a^b r I_m(k_x r) \widehat{K} J_m(k_o r)\, dr - a_{24} \int_a^b r K_m(k_x r) \widehat{K} J_m(k_o r)\, dr \right] \Bigg\} \\ & \times \Bigg\{ \Bigg[ N_m(k_o r) \int_a^b r J_m(k_o r) \widehat{M} K_m(k_x r)\, dr - J_m(k_o r) \int_a^b r J_m(k_o r) \widehat{M} K_m(k_x r)\, dr ] a_{44} \\ & - a_{43} [N_m(k_o b) \int_a^b r J_m(k_o r) \widehat{M} I_m(k_x r) dr - J_m(k_o b) \int_a^b r N_m(k_o r) \widehat{M} I_m(k_x r)\, dr \Bigg] \Bigg\} \end{aligned} \tag{3.44}$$

here the factors, which determine both expressions for $D_x$ and $D(\kappa)$ have the following forms:

$$a_{23} = \left[ k_x K_m'(k_x a) + \frac{m\mu}{a} K_m(k_x a) \right], \; a_{24} = \left[ k_x I_m'(k_x a) + \frac{m\mu}{a} I_m(k_x a) \right] \tag{3.45}$$

$$a_{43} = \left[ k_x K_m'(k_x b) + \frac{m\mu}{b} K_m(k_x b) \right], \; a_{44} = \left[ k_x I_m'(k_x b) + \frac{m\mu}{b} I_m(k_x b) \right] \tag{3.46}$$

As one can see from the analysis of the expression (3.44) the coupling coefficient $D(\kappa) \propto \widehat{M}\widehat{K} \propto \kappa^2$ has a value of the second order of smallness.

Let us discuss in details the structure of the SFXM dispersion relation (see expression (3.43)) in the case of zero external azimuthal magnetic field. The condition $a_{44} = 0$ coincides with the dispersion relation for SFXMs propagating in a cylindrical metal waveguide of radius $b$, which is entirely filled with plasma (see

Sect. 2.2). The condition $a_{23} = 0$ is the dispersion relation of SFXMs propagating along the interface of a metal cylinder with radius $a$, which is immersed into a uniform plasma and is oriented just along the external axial magnetic field.

From the formal (mathematical) point of view, the structure of equation $D_x = 0$ is similar to the structure of the dispersion equation of coupled waves with linear conversion. Thus one can make the following conclusions:

1. $a_{44} = 0$ plays the role of the dispersion relation of that wave, which propagates along the external metal surface of the considered coaxial waveguide;
2. $a_{23} = 0$ plays the role of the dispersion relation of the other wave, which propagates along the internal metal surface of the waveguide;
3. The product $a_{43}a_{24}$ plays the role of the weak coupling parameter between them.

However, it should be underlined here, that the investigated eigen waves cannot interact linearly, because they cannot propagate with the same value of azimuthal mode number $m$ simultaneously. In other words, for a fixed sign of the number $m$ a solution exists either only for the equation $a_{44} = 0$ or only for the equation $a_{23} = 0$. This is the specific property of HF SFXMs: they propagate along a convex metal surface in wide coaxial waveguides only with positive azimuthal mode numbers, $m > 0$, and along a concave metal surface they propagate only with negative mode numbers, $m < 0$.

The eigen frequency of the HF SFXMs, which propagate in a coaxial waveguide entirely filled with cold magneto-active plasma with number $m > 0$, can be approximately calculated from the Eq. (3.43):

$$\omega_x \approx \sqrt{\Omega_e^2 + \omega_e^2 + \frac{m^2c^2}{a^2} + 4\frac{m^2c^2}{a^2}\frac{b-a}{b+a}\exp\left[-2k_x(b-a)\right]}. \quad (3.47)$$

The presence of an external metal surface with a finite radius leads to some increase of the HF SFXM frequency with azimuthal number $m > 0$, propagating in the considered coaxial waveguide, as compared with the frequency of HF SFXMs, propagating around a metal rod immersed into an infinite uniform magneto-active plasma, under the same conditions.

Along an external metal surface, the HF SFXMs propagate in wide coaxial waveguides only with negative values of the mode number $m < 0$. In this case, the value of the SFXM frequency is less than the frequency of HF SFXMs, which propagate in a cylindrical metal waveguide entirely filled with magneto-active plasma for the same other parameters of the waveguide structure. This is evident from comparison of the expressions for the frequencies of HF SFXMs with opposite signs of the azimuthal mode numbers. As it is proved in Chap. 2, for a fixed absolute value of azimuthal mode number $|m|$, the frequency of the flute wave propagating along an internal surface is higher than the frequency of the flute wave propagating along an external surface with radius $r = b > a$ in the opposite direction.

For the case $m < 0$, the expression for the HF SFXM frequency is similar to expression (3.47). There are only two differences, in this case:

1. The fourth summand under the square root is negative;
2. In this summand, one can write $b^2$ instead of $a^2$.

Thus in a coaxial resonator, SFXMs have a different character of propagation and the dependence of their frequency on the effective wave number $k_{ef} = |m|\,\delta/R$ (where $R = a$ or $R = b$, respectively) is different for positive and negative signs of the azimuthal mode number $m$ [14]. As far as HF SFXMs propagate only in wide plasma waveguides (this means that the plasma skin depth is much less than the plasma radius) the presence of the second radial border of the waves' localization weakly influences their dispersion properties. That is why the difference between frequencies of the HF SFXMs propagating in similar non-coaxial structures (see Chap. 2) is also weak.

Concerning practical application of the studied coupled flute modes one can see that the considered coaxial waveguides can be used for designing waveguide phase shifters. Changes in the phases of electromagnetic signals take place on the different interfaces of the coaxial plasma filled resonator in the different directions: with increasing of the azimuthal angle the signals' phase decreases on the external surface $r = b$, and it increases on the internal surface $r = a$. Anyone will be able to select the coaxial waveguide frequency correspondingly to the frequency of the input signal, which can be changed, by means of changing the external axial magnetic field. Application of such type of plasma filled resonators allows one to get a signal with any necessary phase and to take-off simultaneously a set of signals with phase changes assigned in advance.

HF BFOMs propagate in a wide coaxial waveguide with the following angular frequency:

$$\omega_o \approx \sqrt{\Omega_e^2 + \pi^2 c^2 j^2 (b - a)^{-2}}, \quad \text{where } j = 1, 2, 3. \tag{3.48}$$

If the external axial magnetic field and the sizes of the waveguide are large enough, so that the following equality is valid:

$$\frac{\omega_e^2}{c^2} \approx \frac{\pi^2 j^2}{(b - a)^2} - \frac{m^2}{b^2}, \tag{3.49}$$

then the dispersion curves of SFXMs and BFOMs can intersect each other. In this case, their linear conversion becomes possible under the condition of application of an external magnetic field, which has both a large axial component and a small azimuthal component. Therefore frequency corrections to the eigen values of the frequencies of ordinary and extraordinary flute modes can be calculated by the aid of formula (3.5) and their values are linear over the small parameter $\kappa$.

More information can be obtained by numerical studies of the properties of coupled flute modes propagating in coaxial waveguides. In Fig. 3.12, one can see the dependence of the SFW eigen frequency on the effective wave number

**Fig. 3.12** Interaction between SFXMs and three first radial modes of BFOMs propagating in a plasma filled coaxial waveguide for the case: $m = +1$; $\kappa = 0.3$; $Z = 3$; $b = 2a$

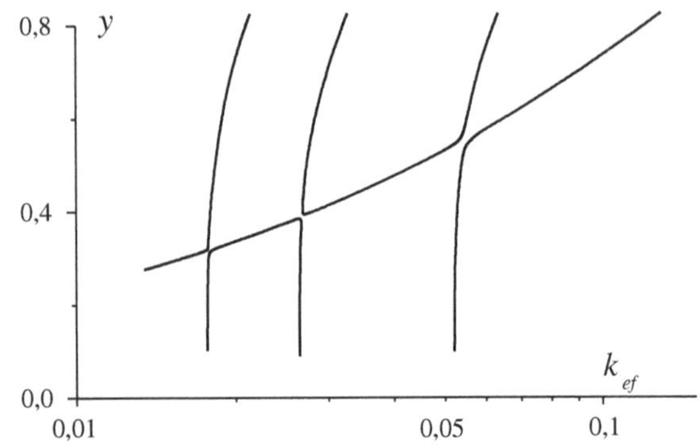


$k_{ef} = |m|\,\delta/b$ for the case with azimuthal mode number $m = +1$. The minimum size of the coaxial line, wherein linear interaction between SFXMs and BFOMs can be realized, is just related to the modes with the minimum value of azimuthal mode number. Increasing of the $Z = \Omega_e/|\omega_e|$ parameter leads to narrowing (in units of electron cyclotron frequency) of the HF range where SFXMs can exist. It should be indicated that this numerical analysis was carried out for the case of linear radial dependence of the external constant azimuthal magnetic field: $B_{0\varphi} \propto r$ and parameter $y$ here has the same meaning as that one, which is applied in Fig. 3.10 (see Sect. 3.3).

In Fig. 3.13, the radial distribution of the axial magnetic and electric fields of SFWs propagating with $m = -1$ in the coaxial waveguide, whose internal radius is four times less than its external radius, is plotted. The following values of the coaxial waveguide parameters were applied for the numerical analysis: $Z = 3$; $k_{ef} = 0.1$; $y = 0.537$. The dashed line depicts the radial dependence of the SFWs' axial magnetic field in zero approximation. It has the typical form for the case of surface wave propagation. This means that the wave amplitude is localized mainly nearby the external border of the considered coaxial waveguide. The bold line shows the radial distribution of the axial electric field of the SFW (it is equal to zero if there is no external azimuthal magnetic field); this dependence is

**Fig. 3.13** Radial distribution of the axial fields for coupled flute modes propagating in a plasma filled coaxial waveguide for the case $m = -1$; $\kappa = 0.3$; $a = b/4$

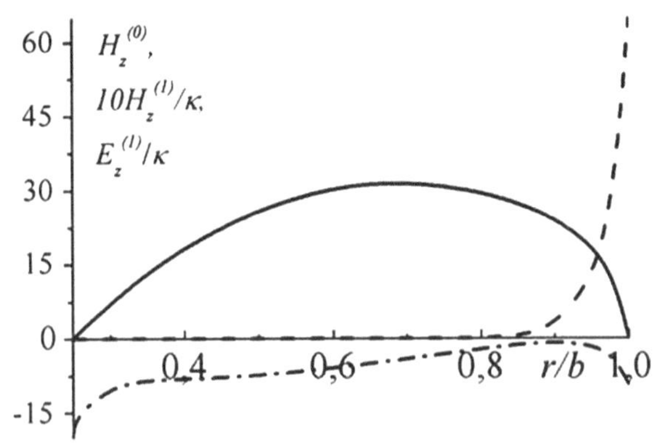

normalized by the small parameter of the problem in order to be clearly visible for the readers. One can see that the shown radial distribution of the axial electric field $E_z^{(1)}$ is the first radial mode of the BFOMs. So in this case the phase difference between the internal and external surfaces of the coaxial waveguide is $k_o(b - a) \approx 2.602$. The dashed-dotted line presents the correction of the radial dependence of the SFW axial magnetic field caused by the presence of an external constant azimuthal magnetic field.

Let us make some common remarks on the results of the numerical analysis carried out in this section. The frequency spectrum of coupled flute modes propagating in coaxial waveguides has some peculiarities compared to the spectra of coupled flute modes studied in the previous sections. First, increasing of the radial mode number of the bulk flute mode results in the fact that the splitting of the frequencies of these flute modes becomes less pronounced nearby the point of their conversion. Second, the range of the $k_{ef}$ values between two neighboring points of their transformation becomes narrower. Third, diminishing of the $k_{ef}$ value leads to a decrease of the SFXM frequency, which means that their frequency approaches to the lower edge of the frequency range (dimensionless parameter $y \to 0$). Fourth, for the SFXMs with $m = +1$, which propagate along the internal metal surface, areas of non-transparency appear (this happens after the second intersection of the dispersion curves plotted in Fig. 3.12, starting from the range of large values of $k_{ef}$). Fifth, the frequencies of BFOMs diminish with increasing distance between the walls of the coaxial waveguide, and intersections of their dispersion curves move toward greater values of $k_{ef}$. Sixth, the minimum size of the coaxial wave-guiding line (wherein linear interaction between SFXMs and BFOMs can be realized) is related just to the flute waves with the minimum value of azimuthal mode number. Seventh, increasing of the $Z = \Omega_e/|\omega_e|$ parameter leads to narrowing (in units of electron cyclotron frequency) of the HF range where SFXMs can exist at all [14].

To estimate the possibility of an experimental observation of this phenomenon let us assume that: plasma density $n = 10^{13}$ cm$^{-3}$, magnetic field $B_{0z} \approx 0.4831$ $T$. Then for realization of linear interaction between the SFXMs and the first radial mode of the BFOMs one has to utilize a coaxial waveguide with $b = 3.12$ cm and $a = 1.56$ cm. In this case [14] the splitting of the frequencies of the SFXM and the BFOM is equal to $2\Delta\omega_X \approx 3 \cdot 10^8$ s$^{-1}$ for the utilized axial electric current of 90 kA.

## 3.5 Conclusions

Results of theoretical investigations on the properties of electromagnetic flute modes propagating in different plasma filled waveguides are represented here for the case that an axial electric current is flowing along the plasma column. The theory of SFWs developed in the Chap. 2 is applied in the third one for studying the linear interaction between the ordinary and extraordinary flute modes in current-carrying plasma waveguides.

The results of studying the propagation of SFWs in current-carrying waveguides with plasma filling seem to be interesting for various practical applications in plasma electronics and technologies. Under different conditions, flowing of an axial electric current leads to coupling between ordinary and extraordinary modes where the forced ordinarily polarized modes can be either of surface or bulk types. In different plasma filled waveguides, coupling between ordinary and extraordinary polarized modes can lead to different phenomena: appearance of non-reciprocal frequency ranges in the frequency spectrum of the coupled flute modes; change of the sign of their group velocity; damping of the extraordinary flute modes caused by their conversion into radiated forced ordinary bulk modes and so on.

The phenomenon of coupling between flute X- and O-modes studied here is linear in respect to temporal evolution of the mutual power transfer of these modes. But power transfer between these modes is characterized by the nonlinear dependence on the small parameter of the problem $\kappa = B_{0\varphi}/B_{0z} \ll 1$.

## References

1. Girka, V., Girka, I., Pavlenko, I., et al. (2012). Coupled azimuthal modes propagating in current-carrying plasma waveguides. *Journal of Plasma Physics, 78*(#2), 105–123.
2. Azarenkov, N. A., & Ostrikov, K. N. (1991). Surface magnetoplasma waves at the interface between a plasma-like medium and a metal in a Voigt geometry. *Physics Reports, 308*, 333–428.
3. Girka, O. I., Girka, I. O., Girka, V. O., et al. (2008). Propagation of azimuthal waves along the surface of a metal current-carrying cylinder immersed into a magnetized plasma. *Radiophysics and Quantum Electronics, 51*(#2), 110–122.
4. Abramowitz, M., Stegun I. A. (1994). *Handbook of mathematical functions.* National Bureau of Standards, Applied Mathematics, Series *55*, New York.
5. Vukovic, V. (1986). *Surface waves in plasmas and solids.* Singapore: World Scientific.
6. Shkaropsky, I. P., Johnston, T. W., & Bachynski, M. P. (1966). *The particle kinetics of plasmas.* London, Ontario: Addison-Wesley Publishing Company.
7. Alexandrov, A. F., Bogdankevich, L. S., & Rukhadze, A. A. (1984). *Principles of plasma electrodynamics.* Berlin and New York: Springer-Verlag.
8. Akhiezer, A. I., Akhiezer, I. A., Polovin, R. V., Sitenko, A. G., & Stepanov, K. N. (1975). *Plasma electrodynamics.* Oxford: Pergamon Press.
9. Krall, N. A., & Trivelpiece, A. W. (1973). *Principles of plasma physics.* NY: MacGraw-Hill.
10. Girka, O. I., Girka, I. O., Girka, V. O., et al. (2008). Propagation of azimuthal waves in magneto-active waveguides filled with a current-carrying plasma. *Technical Physics, 53*(#7), 905–912.
11. Girka, V. O., Girka, I. O. (2002). Effect of toroidal magnetic field variations on the spectra of azimuthal surface waves in metal waveguides entirely filled with plasma. *Plasma Physics Reports, 28*(# 3), 190–195.
12. Lifshits, Ye. M., Pitaevsky, L. P. (1981). *Course of theoretical physics. physical kinetics. 10.* Pergamon Press: Oxford.
13. Landau, L. D., Lifshits, Ye. M. (1960). *Course of theoretical physics. electrodynamics of continuous media. 8.* Pergamon Press: Oxford.
14. Girka, O. I., Girka, I. O., Girka, V. O., et al. (2008). Coupled transverse modes of coaxial metal waveguides completely filled with magneto-active current-carrying plasma. *Plasma Physics Reports, 34*, 901–910.

# Chapter 4
# Surface Flute Waves Propagating in Non-Isotropic Plasma Filled Waveguides

This Chapter is devoted to the theory of surface waves propagating across the axis of symmetry in non-isotropic cylindrical metal waveguides with magneto-active plasma filling. The presented theoretical results illustrate: how can spatial non-uniformity of both the plasma density and the external magnetic field affect the dispersion properties of these waves. These problems have been solved both analytically and numerically. The analytical study of the obtained equations is carried out by the method of successive approximation; adequacy of this approach is proved here as well. Numerical evaluations of the possibilities to experimentally observe the phenomena, which accompany the propagation of these waves in non-uniform plasma filled waveguides, are carried out.

Experimental studies of non-uniform plasma waveguides are important for solving different plasma-physical problems. For instance, the plasma produced during a gas discharge affects the radiation efficiency of the plasma antenna or an array of plasma antennas [1–3] applied this plasma. Anisotropy of a magneto-active plasma strongly affects the spatial distribution of the fields of axially non-symmetric waves, which are emitted by the antenna with such plasma coating. Results of broad theoretical and experimental investigation into excitation of quasi-electrostatic and helicon waves by a ring vibrator immersed into magneto-active plasma are presented in [4]. It is found there, that the radiated power resonantly depends on the plasma density profile. The HF surface wave (SW) radar is shown in [5] as a highly effective device, which can provide over-the horizon surveillance of surface vessels, which move above sea level. From this point of view studying surface flute waves (SFWs) features in non-uniform waveguides seems to be important for the development of new types of powerful antennas.

SW properties are intensively studied for sustaining microwave gas discharges, which have broad practical applications for development of different plasma technologies [6–9]. The parameters of microwave gas discharges depend on different characteristics, such as: geometry of the discharge chamber, type of the operating gas and its pressure, value of the external magnetic field, type of the operating electromagnetic mode and so on. It has been found that utilization of an external magnetic field and axially non-symmetrical SWs allows one to increase the volume of the produced

V. Girka et al., *Surface Flute Waves in Plasmas*, Springer Series on Atomic, Optical, and Plasma Physics 79, DOI: 10.1007/978-3-319-02027-3_4,

plasma and makes the gas discharge longer. Plasma generators based on utilization of transverse waves excited by slot antennas [8, 9] seem to be effective devices because the operating modes, which propagate along the azimuthal direction, do not lose their energy from the discharge chamber. Application of SWs has a lot of advantages as compared with the case of bulk wave application [6]. This is additional reason which determines the importance of studying the properties of SWs, which propagate in non-uniform plasmas.

External magnetic fields, which are usually applied for confining plasma structures in laboratory devices, are also non-uniform. This is connected with the methods utilized for creation of such magnetic fields. To confine a plasma with a small volume one can apply either stationary solid ferromagnetics or a long solenoid. Only for small samples of semiconductor plasmas one can consider that the created magnetic field is uniform. But in the case of gaseous plasma one has to apply a complicated chamber as a trap for the produced plasma and discrete magnetic coils for generation of the confining magnetic field (e.g. toroidal tokomaks and stellarators, mirror magnetic traps, reversed field pinches etc.). Thus any large-scale plasma device has a non-uniform magnetic field. Even if this non-uniformity is small it can affect the properties of the eigen electromagnetic waves, whose frequency is depended upon utilized magnetic field. That is why the problem of non-uniformity of an external magnetic field is under consideration in this Chapter.

## 4.1 Surface Flute Waves in Waveguides Filled with Plasma Whose Density is Non-Uniform

The model of plasmas with uniform density describes in a sufficiently good manner only the case of solid state plasmas [10–13]. Non-uniformity of the plasma density affects the frequencies of both bulk and surface waves. Let us consider a three-component waveguide structure, which consists of a cylindrical metal waveguide of radius $R_2$ which has a dielectric coating in the region $R_2 > r > R_1$ with dielectric permittivity $\varepsilon_d$ and a coaxial plasma cylinder with radius $R_1$ (see Fig. 2.1). The plasma density is supposed to be spatially non-uniform and the value of its density in the central part of the waveguide, which is applied for confining this plasma, is assumed to be large, so that the inequality $\Omega_e^2(r \to 0) \gg \omega_e^2$ is valid, here $\Omega_e$ and $\omega_e$ are the electron plasma and electron cyclotron frequencies, respectively. An external magnetic field is directed along the axial direction $\overrightarrow{B_0} \parallel \overrightarrow{z}$.

Let us study the dispersion properties of SFWs, which depend on coordinates and time in the following form: $E, H \propto f(r)\exp[i(m\varphi - \omega t)]$. Then starting from solving the set of Maxwell equations, one can derive the following equation for the axial magnetic component of the surface flute X-modes (SFXMs) field in the region of non-uniform plasma:

$$\frac{1}{r}\frac{d}{dr}\frac{r}{k_\perp^2}\frac{dH_z}{dr} - \left[1 + \frac{m^2}{k_\perp^2 r^2} - \frac{m}{r}\frac{d}{dr}\left(\frac{\mu}{k_\perp^2}\right)\right]H_z = 0, \tag{4.1}$$

here $\mu = \varepsilon_2/\varepsilon_1$, $k_\perp^2 = k^2(\mu^2 - 1)\varepsilon_1$, $k = \omega/c$, $\varepsilon_j$ are the components ($j = 1, 2$) of the dielectric permittivity tensor for magneto-active plasma in hydrodynamic approximation [10, 14–16], and $m$ is azimuthal mode number.

Applying the method of etalon equations [17] for solving Eq. (4.1) in the plasma region, $r < R_1$ one can derive:

$$H_z = A(r) I_m(z); \tag{4.2}$$

here $A(r) = const \cdot \sqrt{zk_\perp^2/(rz')}$; $z' = k_\perp \sqrt{1 - (m/r)\, d\left(\mu/k_\perp^2\right)/dr}$ and $I_m(z)$ is modified Bessel function [18]. The correctness of this solution depends on the satisfaction of the inequality:

$$1 \gg \left| \frac{m}{r} \frac{d}{dr} \left( \frac{\mu}{k_\perp^2} \right) \right|, \tag{4.3}$$

it describes the case of weak non-uniformity of the plasma density.

The dispersion equation for SFXMs can be derived by the aid of the boundary conditions indicated in Chap. 2 (see Sect. 2.1). In this case the dispersion equation looks like as follows [17]:

$$\frac{m}{R_1}\mu + \frac{1}{H_z}\frac{dH_z}{dR_1} = \frac{k_\perp^2}{k} \frac{J'_m(kR_1)\, N'_m(kR_2) - J'_m(kR_2)\, N'_m(kR_1)}{J'_m(kR_2)\, N_m(kR_1) - J_m(kR_1)\, N'_m(kR_2)}, \tag{4.4}$$

here $J_m(z)$ and $J'_m(z)$ are Bessel functions of the first kind and their derivatives over the argument, respectively; $N_m(z)$ and $N'_m(z)$ are Neumann functions and their derivatives over the argument, respectively [18]. The analysis of Eq. (4.4) allows one to determine the frequency ranges where SFXMs can propagate. There are two possible ranges, which can be referred here in the same manner as it is done in the Sect. 2.1, namely the LF range (2.3) and the HF range (2.4), respectively. Applying asymptotes [18] of the Bessel functions in the limiting case $|m| \gg 1$, one can obtain analytical expressions for the SFXMs frequency in the HF and LF ranges, respectively:

$$\omega \approx \omega_1 \left[ 1 + \frac{\delta^2}{2R_1^2} \left( \Delta \frac{R_1^2}{\delta^2} \frac{|\omega_e|}{\omega_1} + \frac{m\Omega_e^2}{\omega_1^2} - \frac{\Omega_e^2 \xi}{\omega_1 |\omega_e|} \right)^2 \right], \tag{4.5}$$

$$\omega \approx \frac{|\omega_e|}{\sqrt{1 + k_0^{-2}}} \left[ 1 - \Delta \frac{\Omega_e^2}{\omega_e^2} |m| \sqrt{1 + k_0^{-2}} - \frac{\xi R_1^2 \omega_e^2}{m^3 c^2 (1 + k_0^{-2})^{3/2}} \right]^{-1}, \tag{4.6}$$

here $\omega_{1,2} = \mp 0.5\, |\omega_e| + \sqrt{\Omega_e^2 + 4^{-1}\omega_e^2}$ are the cut-off frequencies, $\omega_{LH}$ and $\omega_{UH}$ are the lower hybrid and upper hybrid frequencies, respectively, $\delta = c/\Omega_e$, $k_0 = m\delta R_1^{-1}$, $\xi = (R_1\delta^{-1})(d\delta/dR_1)$ is a parameter which describes the non-uniformity of the plasma density, and $\Delta = R_2 R_1^{-1} - 1$ is the dimensionless thickness of the dielectric layer, which separates the non-uniform plasma column from the metal wall of the waveguide.

Using solutions (4.5) and (4.6) it is possible to obtain approximate expressions for the angular frequency $\omega$ of SFXMs in other limiting cases, namely: $z(R_1) \ll 1$ and $z(R_1) \gg |m|$. The results of a numerical analysis of dispersion Eq. (4.4) are presented in [17]. There it was found that the condition of weak non-uniformity of the plasma density slightly changes the SFXM frequency as compared with the case of uniform plasma. Since these extraordinary polarized modes have a non-reciprocal frequency spectrum (propagate with different frequencies in mutually opposite directions, under the same conditions), increasing radial plasma density gradient $d\Omega_e^2/dR_1$ affects the SFXMs frequency in the cases of positive and negative signs of the azimuthal mode numbers $m$ in different way. For instance, the angular frequency of LF SFXMs with $m < 0$ diminishes more slightly as compared with the case of positive azimuthal mode numbers $m > 0$.

It can be pointed out that the dispersion Eq. (4.4) once more describes two waveguide structures, namely a waveguide, which is partially filled with a cold non-magnetized plasma ($B_0 = 0$) and a waveguide which is entirely ($\Delta = 0$) filled with a cold magneto-active plasma. In the first case one can obtain the expression for the angular frequency of SFXMs within an accuracy up to the summands of the second order of smallness, $\Delta^2 \ll 1$:

$$\omega^2 \approx \frac{\Delta\left(1 - \Delta/2\right) m^2 \Omega_e^2}{\sqrt{m^2 + R_1^2/\delta^2 + \Delta\left(1 + \Delta/2\right) R_1^2/\delta^2 - \xi}}. \tag{4.7}$$

In the second case, it is possible to derive the SFXM frequencies just from expressions (4.5) and (4.6) putting there $\Delta = 0$. But it can be underlined that if $\Delta = 0$ then SFXMs become unidirectional modes (i.e. they can propagate only along unique direction); thus only SFXMs with $m > 0$ can propagate in the LF range $\omega < \omega_1$ and only a negative sign of the azimuthal mode number $m < 0$ can be realized for SFXMs in the HF range $\omega > \omega_{UH}$. Therefore, analyzing the expressions (4.6) and (4.7) one can make the conclusion: if the plasma density increases in the direction towards the plasma border then the SFXM frequency value diminishes as compared with the case of uniform plasma density profile.

In the case of arbitrary type of the plasma density profile, there is a unique method to determine the SFXM spectrum; it is a numerical analysis. But there is a special limiting case, if SFXMs propagate in a wide plasma cylinder (plasma radius is considerably larger than the SFXM penetration depth into the plasma) then their property can be studied analytically. It can be shown that if nearby the plasma boundary its density is characterized by a linear profile then dispersion Eq. (4.4) can be solved analytically.

Let us restrict our further considerations in this section only to the case of LF SFXMs propagating in a metal waveguide filled with cold magneto-active plasma.

It is assumed that the plasma, which is located in the region $r \le R_1$, is characterized by the following density profile [19]:

$$n(r) = n(R_1) + (r - R_1)dn/dr\left.\right|_{r=R_1} . \tag{4.8}$$

In the case of a wide plasma cylinder ($R_1 \gg |m|\,\delta_1$, here $\delta_1 = c\Omega_e^{-1}(R_1)$) the inequality $\left|m^2k_\perp^{-2} - (m/r)d(qk_\perp^{-2})/dr\right| \ll 1$ is valid. Then applying the representation of the SFXM field as a series: $H_z = H_z^{(0)} + H_z^{(1)} + \ldots$, where $\left|H_z^{(n+1)}\right| \ll \left|H_z^{(n)}\right|$, one can solve Eq. (4.1) by the aid of the method of successive approximations. Let us write down here the expressions for the zero and first order approximations, respectively:

$$H_z^{(0)} = C_1U'(a) + C_2V'(a), \tag{4.9}$$

$$H_z^{(1)} = U'(a)\int\limits_{a_1}^{a} V'(a)g da - V'(a)\int\limits_{a_2}^{a} U(a)g da. \tag{4.10}$$

Here $g = -\frac{\delta_0^2}{r}\frac{dH_z^{(0)}}{dr} + q(r)H_z^{(0)} - \frac{k^2}{r}\frac{d}{dr}\left[\delta_0^4 r\left(1+\frac{\omega_e^2}{\omega^2}\right)\frac{dH_z^{(0)}}{dr}\right]$, $\left|\frac{g}{H_z^{(0)}}\right| << 1$,
$\delta_0 = c/\Omega_e(r < R_1)$, $a(r) = (\delta_1 k_0)^{-2} + k_0(R_1 - r)$, $k_0^3 = -d\delta_0^{-2}/dr\left.\right|_{R_1}$, $U'(a)$ and $V'(a)$ are derivatives of Airy functions over their arguments [18]. Applying the asymptotes of the Airy function for large arguments and choosing the solution of Eq. (4.1) in the form of SWs, one finds that $C_1 = 0$, $a_1 = a(0)$ and $a_2 = a(R_1)$ if the plasma density value increases along the direction from plasma border towards waveguide wall (this means that $k_0 > 0$). In the opposite case, if $k_0 < 0$ then $C_2 = 0$, $a_1 = a(R_1)$ and $a_2 = a(0)$. To find solutions of Eq. (4.1) in the form of the expressions (4.9) and (4.10) it is not sufficient to satisfy the inequalities $R_1 \gg |m|\,\delta$ and $\Omega_e^2 \gg \omega_e^2$, however, it is necessary also to provide a smooth variation of the plasma density, so that the following inequality can be realized:

$$|\delta_0 d\;\ln[n(r)]/dr|_{|r=R_1} \ll 1. \tag{4.11}$$

Condition (4.11) is satisfied if the inequality $k_0^3 \le (R_1\delta_1^2)^{-1}$ is valid. In this case the dispersion equation for SFXMs looks like in the previous case (see Eq. (4.4)).

Let us substitute solutions of Maxwell equations in the forms (4.9) and (4.10) into Eq. (4.4) and apply asymptotes of the Airy functions for large argument ($a \gg 1$) [18]. Then one can derive the following expression for LF SFXMs in approximation of a narrow vacuum layer $k(R_2 - R_1) \ll 1$:

$$\omega \approx \frac{0.5m\,|\omega_e|}{R_1/\delta_2 + \Delta R_1^2/\delta_2^2} + \sqrt{\frac{0.25m^2\omega_e^2}{(R_1/\delta + \Delta R_1^2/\delta^2)^2} + \frac{\Delta m^2\Omega_e^2(R_1)}{R_1/\delta + \Delta R_1^2/\delta^2}}, \tag{4.12}$$

here $\delta_2 = \delta_1(1 + k_0^3\delta_1^3/4)$.

Summarizing the obtained above results one can make the following conclusions. First, application of the approximation of the plasma density profile by a linear function allows one to solve the problem of the influence of plasma density non-uniformity on the LF SFXM frequency spectrum analytically. Second, the type of plasma density profile can be a nonlinear one with respect to the plasma depth, for example, it can change from a linear form nearby the plasma periphery into a parabolic form in the plasma interior, but such a behavior will not affect the obtained solution. Third, if the skin depth is $\delta = \delta_1$, formula (4.12) coincides with the corresponding expression, which was obtained for the case of uniform plasma in Sect. 2.1, (see Eq. (2.11)). Fourth, analyzing (4.12) one can see that if the density of the plasma decreases along the direction from the plasma boundary towards the waveguide wall, the effective $\delta$ penetration depth of the LF SFXMs into the plasma is less than the corresponding value in the case of a uniform plasma. Fifth, in the case of the form of the density profile indicated above, the angular frequency of LF SFXM becomes less than it was in the case of a uniform plasma as well.

Let us consider another example of a plasma density profile, which allows one to solve the problem of changing SFW frequency analytically. Studying the tasks of propagation of electromagnetic waves in thermonuclear fusion devices filled with gaseous plasma, the following form of radial dependence for the plasma density nearby the boundary of the plasma is often applied:

$$n(r) = (r - R_1)dn/dr\left|_{r=R_1}\right. . \tag{4.13}$$

Here the gradient of the plasma density is assumed to have the form: $\left|dn/dr\right|_{|r=R_1} \sim n(0)R_1^{-1}$. The plasma density reaches such a value nearby the axis of the studied waveguide, so that the inequality $|\varepsilon_1(0)| \gg 1$ becomes valid. Then for the plasma region the following inequality $R_1 - r \gg k^2 k_0^{-3}$ will be satisfied. So it is possible to apply the solution of Eq. (4.1) in the form of Eqs. (4.9) and (4.10) using the following replacement $a \to k_0(R_1 - r)$, if the following inequalities: $k_0^2 \gg k^2$, $k_0 R_1 \gg |m\omega_e|/\omega$, $R_1 \gg k_0^{-1}$ are satisfied. After that in the peripheral plasma region (nearby plasmas interface $0 \le R_1 - r \ll k_0^{-1}$), the solution of Eq. (4.1) can be found with the aid of the method of a narrow layer [19]:

$$H_z = H_z(R_1) + H_z(R_1)\int\limits_{R_1}^{r} dr' k_\perp^2 \left[A_d + \int\limits_{R_1}^{r'} dr \frac{H_z(r)}{H_z(R_1)}(1 + q(r))\right] . \tag{4.14}$$

The constant of integration $A_d$ can be determined with the aid of application the boundary conditions on the plasma-vacuum interface, which are considered in details in Chap. 2:

$$A_d = \frac{1}{k}\frac{N_m'(kR_1)J_m'(kR_2) - N_m'(kR_2)J_m'(kR_1)}{N_m'(kR_2)J_m(kR_1) - N_m(kR_1)J_m'(kR_2)} . \tag{4.15}$$

In the case of a narrow dielectric layer $R_2 - R_1 \ll R_1$ one can calculate an approximate value of the constant $A_d$ using an expansion procedure over the small parameter $\Delta \ll 1 : A_d \propto \Delta$.

The expressions for the SFXM fields obtained in the form of (4.9) and (4.10) in the region $R_1 - r \gg k^2 k_0^{-3}$ and those obtained in the form (4.14) in the region $k^2 k_0^{-3} \leq R_1 - r \ll k_0^{-1}$ can be joined. This allows one to derive a dispersion equation, whose solution in the considered case (plasma density profile is linear nearby the plasma border) has the following form:

$$\omega \approx \frac{\sqrt{9k_0^4 \Delta m^2 c^2 (\Delta R_1^2 + V_1 R_1 k_0^{-1}) + V_1^2 m^2 \omega_e^2}}{3k_0^2 (\Delta R_1^2 + V_1 R_1 k_0^{-1})} + \frac{V_1 m \left|\omega_e\right|}{3k_0^2} \left( \Delta R_1^2 + \frac{V_1 R_1}{k_0} \right)^{-1}, \tag{4.16}$$

here $V_1 = -V(0)/V'(0) \approx 1.37$. From comparison of the expressions (4.12) and (4.16) one can make the conclusion that in the case of a linear density profile the obtained expression for the SFXM frequency as function of the plasma parameters is similar to that one, which was obtained in the case of a uniform plasma when doing the replacement of the SFXM penetration depth $\delta_1$ by the parameter $k_0^{-1}$ [20].

In this section, the frequency $\nu$ of collisions between plasma particles has not been taken into the account so far, but under experimental conditions its value can be often not so small. That is why one should take into account the collision frequency $\nu$ in the expressions for the components of the dielectric permittivity of the plasma which leads to the appearance of SFXM damping. The damping rate of SFXMs determined by plasma particle collisions can be calculated from dispersion Eq. (4.4), which describes the case of a non-uniform plasma. Making the replacement $\omega \to \omega - i\gamma_c$, the following expression for the collisional damping rate can be obtained:

$$\gamma_c \approx \frac{\nu}{3} \left[ 1 + \frac{2m\left|\omega_e\right|}{3R_1 k_0 \omega} \right] \left[ \frac{2m\left|\omega_e\right|}{3R_1 k_0 \omega} + \frac{2k_0 m^2 \Delta}{V_1 k^2 R_1} \right]^{-1}. \tag{4.17}$$

Expression (4.17) in the limiting case $R_1 \to \infty$ turns to $\gamma_c \approx \nu/3$, which is typical for a planar plasma-vacuum border [11, 21]. Analyzing expression (4.17) one can make the conclusion that the collisional damping rate of SFXMs propagating along a plasma-metal boundary is larger than that one calculated in the case of a plasma-dielectric boundary. It is possible to explain such behavior of the SFXMs by the circumstance that in the first case the SFXM energy propagates only in the plasma region, while in the second case the wave energy propagates as well in the region of the dielectric layer, which separates the plasma column from the waveguide metal wall.

There are some more reasons for SFXM damping in the case of their propagation in non-uniform plasma. If the SFXM frequency exceeds the electron cyclotron frequency, the resonance point $\mathrm{Re}(\varepsilon_1) = 0$ can be located nearby the plasma border. This leads to substantial increasing of the SFXM damping. In this case, the

SFXM damping rate consists of two terms, namely the collisional and resonant damping rates $\gamma_c + \gamma_r$. To determine the resonant damping rate $\gamma_r$, whose appearance is connected with the presence of the resonance point one can take into the account the imaginary part of the component $\varepsilon_1$ of the dielectric permittivity tensor of the plasma filling in the parameter $k_\perp^2$, which enters into the first summand of Eq. (4.1). It will allow one to calculate the value of the resonant damping rate in the following form:

$$\gamma_r \approx \frac{\pi\omega\, m^2 k_0}{\varsigma\, k^2 R_1^2}\left(1 - \frac{m\,|\omega_e|}{3R_1\omega\, k_0}\right)\left[V_1 + \frac{\Delta k_0}{k^2 R_1}\left(k^2 R_1^2 + m^2 - \frac{2m^3\,|\omega_e|}{3R_1\omega\, k_0}\right)\right]^{-1}, \tag{4.18}$$

where $\varsigma = \left(d\varepsilon_1/dr\right)_{|r=R_1} > 0$. Taking into the account the expression for $k_0$, one can conclude that the value of the SFW resonant damping rate is proportional: $\gamma_r \propto (dr/d\varepsilon_1)^{2/3}$. In the case of SFWs propagating in a waveguide immersed into a weak external azimuthal magnetic field $\vec{B}_0 = B_{0\varphi}\vec{e}_\varphi$ (see Sect. 3.1), the resonant damping rate value depends on the gradient of the plasma density in a stronger way, but there was as well a squared small parameter of that problem ($\beta = |\omega_e|/\omega$), so its value was proportional: $\gamma_r \propto \beta^2 (dr/d\varepsilon_1)^{5/3}$ (in detail, see expression (3.8)).

Expression (4.18) is also valid for the cases of non-magnetized plasma and SFW propagation along the plasma-metal interface. Its analysis confirms that resonant damping of SFXMs becomes larger with moving the resonant point from the plasma border towards its interior and with moving plasma surface towards the metal wall of the waveguide. This corresponds to the results, which were obtained for other types of SWs [11, 22].

Thus, if the plasma density is characterized by a linear profile nearby the plasma periphery the dispersion properties of SFXMs are similar to those determined for the case of a uniform plasma. The influence of the plasma non-uniformity can be taken into account with the aid of replacement of the SFXM penetration depth into the plasma by another definite parameter. Expressions for the SFXM damping rates correspond to the results obtained for the other SWs. Their values increase with increasing value of the ratio $|m|\,R_1^{-1}$.

Let us illustrate possibilities to apply a numerical method to study the dispersion properties of SFWs using an analytical expression of the radial plasma density profile, which is calculated in [23] for the conditions of a HELIAS reactor. As far as the value of the applied external magnetic field in this case is substantially large ($2 \div 4.5$ T), then SFWs can propagate in the LF range ($\omega < |\omega_e|$), moreover their frequency is close to harmonics of the ion cyclotron frequency.

Let us consider the plasma region as a composition of N uniform plasma cylindrical layers, which have different thickness and density values. The inner (first) plasma layer has the density $n_I$ and it is located in the region $0 \le r \le r_1$. The second plasma layer with density $n_2 < n_I$ is located in the region $r_1 < r < r_2$. The other plasma layers with densities: $n_3, n_4, \ldots, n_N$ are located in the following regions, $r_2 < r < r_3$, $r_3 < r < r_4, \ldots, r_{N-1} < r < r_N$, respectively. A realistic

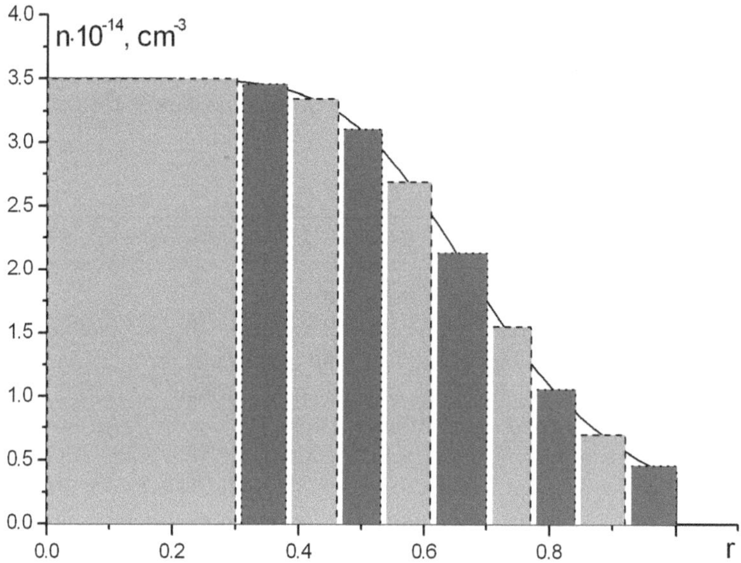


**Fig. 4.1** Dependence of the plasma density on the radius of the plasma column normalized by the plasma radius (the analytical dependence [23] is drawn by the solid line and the modeled profile is represented by *dashed lines*)

plasma density profile calculated according to the results obtained in paper [23] and its model representation in the form of a set of uniform layers are drawn in the Fig. 4.1.

In all the plasma layers the magnetic component of the SFXMs is described by Eq. (4.1) where one can assume that $\frac{\partial}{\partial r}\left(\frac{\mu}{k_\perp^2}\right) = 0$. Thus it becomes a form of modified Bessel equation. Therefore the $H_z$ component of the SFXM field is described by a superposition of modified Bessel functions and MacDonald functions [18]. But taking into account that the SFWs fields have restricted magnitudes on the axis of the plasma column one can equate the MacDonald functions' multiplier to zero for the first plasma layer with density $n_1$. The expression for the tangential component of the SFXM electric field can be obtained from the set of Maxwell equations (see expression (2.12)).

After calculation explicit expressions for $H_z$ and $E_\varphi$ it is easy to find the impedance $Z_{pl}^{(1)}(r_1) = E_\varphi(r_1)/H_z(r_1)$ of the first plasma layer on its interface $r = r_1$:

$$Z_{pl}^{(1)}(r_1) = \frac{iI_m'(k_{\perp 1} r_1)}{\psi_1 I_m(k_{\perp 1} r_1)} + \frac{im\mu_1}{k_{\perp 1} r_1 \psi_1^2}, \tag{4.19}$$

here the prime superscript means the derivative of the modified Bessel function with respect to its argument, subscript 1 means that the corresponding parameter is calculated just for the first plasma layer, namely $\mu_1 = \varepsilon_2(n_1)/\varepsilon_1(n_1)$,

$\psi_1 = \psi(n_1) = \sqrt{(\mu_1^2 - 1)/\varepsilon_1(n_1)}$, $k_{\perp 1} = k_\perp(n_1)$. As far as MacDonald and modified Bessel functions have restricted values in all other plasma layers, both $H_z$ and $E_\varphi$ are expressed by both $I_n(z)$ and $K_n(z)$. Thus the impedances for the rest of the plasma layers can be written in the similar manner:

$$Z_{pl}^{(j)}(r) = \frac{i[A_1^{(j)} I'_m(k_{\perp j} r) + A_2^{(j)} K'_m(k_{\perp j} r)]}{\psi_j [A_1^{(j)} I_m(k_{\perp j} r) + A_2^{(j)} K_m(k_{\perp j} r)]} + \frac{im\mu_j}{k_{\perp j} r \psi_j^2}, \tag{4.20}$$

here $j = 2, 3, \ldots, N$ is the number of the plasma layer.

As it has been mentioned above the SFW fields in the vacuum layer are described by Bessel functions of the first kind and Neumann functions. So in the vacuum region, one can apply Eq. (2.2) which was found in Sect. 2.1 for $H_z^{vac}$. Then $E_\varphi^{vac}$ can be expressed in the following way: $E_\varphi^{vac} = dH_z^{vac}/d(ikr)$. There are two boundary conditions for the impedance $Z_v(r) = E_\varphi^{vac}(r)/H_z^{vac}(r)$ of the vacuum layer: first, it is equal to zero on the metal wall $Z_v(R_2) = 0$; second, there is continuity of the plasma impedance and vacuum impedance on the boundary between plasma column and vacuum layer $Z_{pl}^{(N)}(R_1) = Z_v(R_1)$. So one can obtain the following expression for $Z_v(R_1)$:

$$Z_v(R_1) = \frac{-i[J'_m(kR_2)N'_m(kR_1) - J'_m(kR_1)N'_m(kR_2)]}{J_m(kR_1)N'_m(kR_2) - J'_m(kR_2)N_m(kR_1)}. \tag{4.21}$$

It should be noted that the relative value of the vacuum layer thickness is small, so that inequality $\Delta \ll 1$ is satisfied here.

After that one can apply the boundary conditions for $Z_{pl}^{(j)}(r)$ on the interfaces between all neighbouring plasma layers. This allows one to calculate the impedances for the neighbouring plasma layers on the mutual borders. Doing in such way one can find the following set of $N + 1$ equations for the coefficients $B_j$, which can be considered as equivalent of the SFXM dispersion equation for the case of a non-uniform plasma density profile [24]:

$$B_1 \left\{ \psi_1 \psi_2 I'_m(r_1 k_{\perp 2}) - I_m(r_1 k_{\perp 2}) \left[ \frac{\psi_2^2 I'_m(r_1 k_{\perp 1})}{I_m(r_1 k_{\perp 1})} + \frac{m(\mu_1 \psi_2^2 - \mu_2 \psi_1^2)}{r_1 k_{\perp 1}} \right] \right\} =$$

$$K_m(r_1 k_{\perp 2}) \left[ \frac{\psi_2^2 I'_m(r_1 k_{\perp 1})}{I_m(r_1 k_{\perp 1})} + \frac{m(\mu_1 \psi_2^2 - \mu_2 \psi_1^2)}{r_1 k_{\perp 1}} \right] - K'_m(r_1 k_{\perp 2}) \psi_1 \psi_2,$$

$$B_N \left\{ I'_m(k_{\perp N} R_1) + \psi_N I_m(k_{\perp N} R_1) \left[ \frac{J'_m(kR_1)N'_m(kR_2) - J'_m(kR_2)N'_m(kR_1)}{J_m(kR_1)N'_m(kR_2) - J'_m(kR_2)N_m(kR_1)} + \frac{m\mu_N}{k_{\perp N} R_1 \psi_N} \right] \right\} =$$

$$-K'_m(k_{\perp N} R_1) - \psi_N K_m(k_{\perp N} R_1) \left[ \frac{J'_m(kR_1)N'_m(kR_2) - J'_m(kR_2)N'_m(kR_1)}{J_m(kR_1)N'_m(kR_2) - J'_m(kR_2)N_m(kR_1)} + \frac{m\mu_N}{k_{\perp N} R_1 \psi_N} \right],$$

$$\frac{B_{j-1}I'_m(r_{j-1}k_{\perp j-1})+K'_m(r_{j-1}k_{\perp j-1})}{B_{j-1}I_m(r_{j-1}k_{\perp j-1})+K_m(r_{j-1}k_{\perp j-1})}+\frac{m(\mu_{j-1}\psi_j^2-\mu_j\psi_{j-1}^2)}{\psi_j^2 r_1 k_{\perp j-1}}= \tag{4.22}$$

$$\frac{B_j I'_m(r_{j-1}k_{\perp j})+K'_m(r_{j-1}k_{\perp j})}{B_j I_m(r_{j-1}k_{\perp j})+K_m(r_{j-1}k_{\perp j})}\frac{\psi_{j-1}}{\psi_{j+1}},$$

where the subscript $j$ in the last equation of the set (4.22) changes in the limits from 2 to $N$. Thus the last equation really represents $N-1$ equations obtained from equating the impedances of the neighboring plasma layers beginning from the impedance of the second layer.

The results of a numerical analysis of the set of Eqs. (4.22) are presented in Figs. 4.2, 4.3 and 4.4. The value of the SFXM frequency decreases with increasing plasma radius more strongly as compared with that one calculated in the case of a uniform plasma (see Fig. 4.2). Increasing the vacuum layer thickness leads to decreasing SFXM frequencies, under the same conditions. The results of this numerical analysis prove that their frequency value decreases with increasing confining magnetic field value (see Fig. 4.3).

It is found that in the case of a non-uniform plasma density profile the decrease of the SFXM frequency value with increasing plasma column radius happens faster than in the case of a uniform plasma. So there is a cross point of

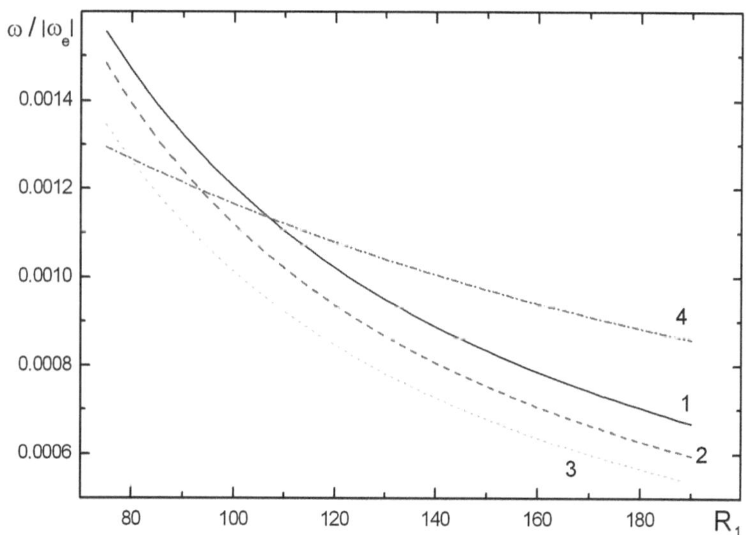


**Fig. 4.2** Frequency of the SFXMs versus plasma column radius (in cm) for $m=3, B_0=3T$. The numbers 1, 2 and 3 relate to the cases $\Delta=0.05$; 0.25 and 0.45, the line marked by 4 relates to the case a of a uniform plasma, $\Delta=0.25$

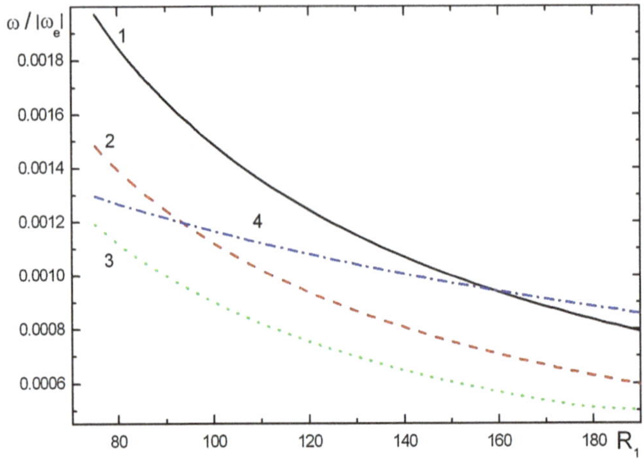


**Fig. 4.3** The same as in Fig. 4.2 for $m = 3, \Delta = 0.25$. The numbers *1*, *2* and *3* relate to the cases $B_0 = 3$; 4 and 5 *T*, respectively. The line marked by *4* relates to the case of a uniform plasma, $B_0 = 4\ T$

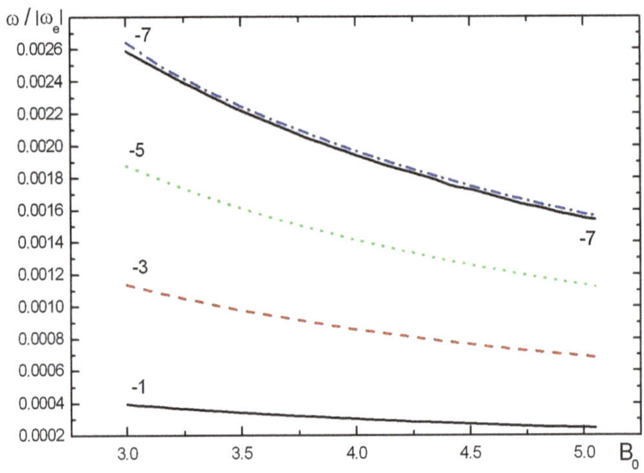


**Fig. 4.4** Frequency of the SFXMs versus confining magnetic field (in *Tesla*) for $R_1 = 130$ cm, $\Delta = 0.25$. The numbers indicate the values of the azimuthal mode number $m$. The *solid line* marked by $-7$ relates to the case of a uniform plasma

the dispersion curves related to the cases of uniform and non-uniform plasmas. The results of a numerical analysis of the SFXM frequency dependence on the azimuthal mode number $m$ confirm the conclusion obtained analytically (see Fig. 4.4). It is confirmed numerically that there are two possible propagation directions of SFXMs with positive and negative values of the azimuthal mode number (see Chap. 2). It is found that $\omega(+m) < \omega(-m)$, under the same conditions, but the relative difference between these frequencies values is small. For instance, in the case of a uniform plasma, the magnitude of order of this relative difference is ~$10^{-2}$ and in the case of a non-uniform plasma, it is ~$10^{-1}$. Features of SFXMs with opposite values of azimuthal mode numbers are similar to each other therefore figures related to the case of $m < 0$ are not presented.

Analyzing the results presented in Fig. 4.4 one can see that non-uniformity of the plasma density leads to some increase of the SFXM frequency as compared with the case of uniform strongly magnetized plasma. Thus numerical investigations confirm the conclusions of the analytical studies on the increase of the value of the SFXM frequency with decreasing values of confining magnetic field, plasma radius and thickness of the vacuum layer.

## 4.2 Influence of Radial Non-Uniformity of External Magnetic Field on Surface Flute Waves Frequency Spectrum

Plasma confinement in various magnetic traps is performed by the aid of an external magnetic field $\vec{B}_0(r)$ with non-uniform radial profile. As a rule this non-uniformity is relatively weak. That is why in this section just the case of a weak radial non-uniformity of the magnetic field and its influence on the SFXM frequency spectrum are considered. As far as the influence of the non-uniformity of the plasma density on SFXMs properties has been examined in the previous section, let us assume here the case of a uniform plasma density profile.

It is interesting that in this case the magnetic component of the extraordinary polarized SFXM field is also described by Eq. (4.1). Because of this it can be solved by using the method of etalon equations [17]. Applying this method one can find that in the plasma region, the axial magnetic field of SFXMs is described by the following expression:

$$H_z = A(r)I_m(b). \tag{4.23}$$

This solution is valid if condition (4.3) is satisfied. Substituting the expression (4.23) into Eq. (4.1) one can calculate the correction $\omega_1$ to the SFXM frequency, which is connected with radial non-uniformity of an external stationary magnetic field $\vec{B}_0$ in the limiting case of a wide plasma waveguide $R_1 \gg |m|\,\delta_1$ and a large azimuthal mode number $|m| \gg 1$:

$$\omega_1 \approx \frac{-m\xi\omega^2}{2\sqrt{m^2 + R_1^2/\delta_1^2}}\left[m\omega + 2\Delta\frac{m^2\Omega_e^2}{|\omega_e(R_1)|}\right]^{-1}. \tag{4.24}$$

Here the characteristic parameter of the radial non-uniformity of the external magnetic field $\xi$ is determined by the expression $\xi = R_1 d\ \ln B_0/dR_1$, which is calculated in the point of $r = R_1$, $\delta_1 = c/\Omega_e(R_1)$. Under the considered conditions the following inequality for the parameter of the magnetic field non-uniformity is valid $\xi \ll R_1^2\omega / \left|m\omega_e(R_1)\delta_1^2\right|$. Expression (4.24) can be applied as well for obtaining the frequency corrections $\omega_1$ of these modes in other limiting cases, namely narrow plasma waveguide $k_\perp R_1 \ll 1$ and small azimuthal mode number $|m| \ll k_\perp R_1$.

If the thickness of the dielectric layer is sufficiently small, the inequality: $\Delta < |\omega_e(R_1)|\,\omega / \left|m\Omega_e^2\right|$ is valid and the SFXMs become unidirectional waves (in this case the azimuthal mode number can be only positive: $m > 0$). In an external magnetic field, whose strength decreases along the direction from the plasma boundary, their frequency is less than in the case of a uniform magnetic field. This conclusion agrees with the results of Ref. [17] concerning the SFXM frequency as function of the applied magnetic field $B_0$ magnitude. If the thickness of the dielectric layer, which separates the confined plasma column from the metal wall of the waveguide, is so large that the following inequality $\Delta > |\omega_e(R_1)|\,\omega / \left|m\Omega_e^2\right|$ is valid, the influence of an external magnetic field is more pronounced for SFXMs with negative sign of the azimuthal mode number $m < 0$, compared with the case of positive sign. This can be explained as demonstration of the appearance of a non-reciprocal frequency spectrum for SFXMs (this means that in the case of propagation of these waves in opposite directions their frequencies are different, under the same conditions).

Summarizing the results presented above, one can make the following conclusions. First, radial non-uniformity of an external magnetic field affects much stronger those SFXMs, which propagate along the direction of electron rotation in the magnetic field $B_0$ nearby the plasma interface. Second, for fixed value of the characteristic parameter $\xi$ of the external magnetic field radial non-uniformity, the correction of the SFXM frequency decreases with increasing thickness of the dielectric layer, absolute value of the azimuthal mode number, plasma radius, plasma density in the waveguide and value of the applied magnetic field $B_0(R_1)$.

## 4.3 Influence of Toroidal Magnetic Field Non-Uniformity on Surface Flute Waves Frequency Spectrum in Metal Waveguides Entirely Filled with Plasma

In a gaseous plasma, which is confined in a toroidal experimental device, the influence of the toroidicity of the utilized magnetic field on the eigen transverse electromagnetic waves is revealed at first, via a spatial non-uniformity of the constant external toroidal magnetic field; at second, via a toroidal shift of the magnetic surfaces in these experimental devices; and at third, via deviation of the shape of the poloidal cross-sections of the magnetic surfaces from circular cross-section. The influence of the second and third factors on the dispersion properties of the eigen electromagnetic waves, which propagate across the uniform axial magnetic field along a small azimuthal angle is particularly examined in [25].

This section is devoted to the study of the influence of the first feature mentioned above on the properties of SFWs, which propagate in a cylindrical waveguide entirely filled with plasma. Doing that the method of successive approximations is applied. In the frame of this theory, the SFXMs propagation in a uniform magnetic field is applied as zero approximation. As it is shown in the previous section an external stationary magnetic field very strongly affects the SFXM dispersion properties in the case of the absence of a dielectric layer, which separates the plasma column from the metal wall of the utilized waveguide. That is why we just examine the case of a toroidal metal waveguide with circular cross-section and radius $a$, which is entirely filled with cold plasma.

To solve the problem one can choose the suitable co-ordinate system. In the right-hand quasi-toroidal co-ordinate system (it means that the vector product of $\vec{e}_r$ with $\vec{e}_\vartheta$ is equal to $\vec{e}_\zeta$), the chamber is assumed to be symmetric along the circular axis, so $\partial/\partial\zeta = 0$. In this case the poloidal angle $\vartheta$ is counted from the direction of the center of symmetry of the torus (see Fig. 4.5, where this co-ordinate system is also shown). It is assumed as well that the plasma filling is uniform and the applied external constant toroidal magnetic field is described by the following expression:

$$B_{0\zeta} = B_0 / \left[1 - \left(r/R\right)\cos\vartheta\right], \tag{4.25}$$

here $R$ is the major radius of the torus. As it will be shown here, eigen extraordinary polarized SFWs with the components of the electromagnetic field $E_r$, $E_\vartheta$ , $H_\zeta$ can propagate in such toroidal waveguides. In electrodynamics of restricted plasma systems, propagation of SWs across an external stationary magnetic field oriented parallel to the plasma-metal interface is entitled wave propagation in Voight geometry [12, 22].

**Fig. 4.5** Scheme of the waveguide structure consisting of a toroidal metal ring immersed into magneto-active plasma

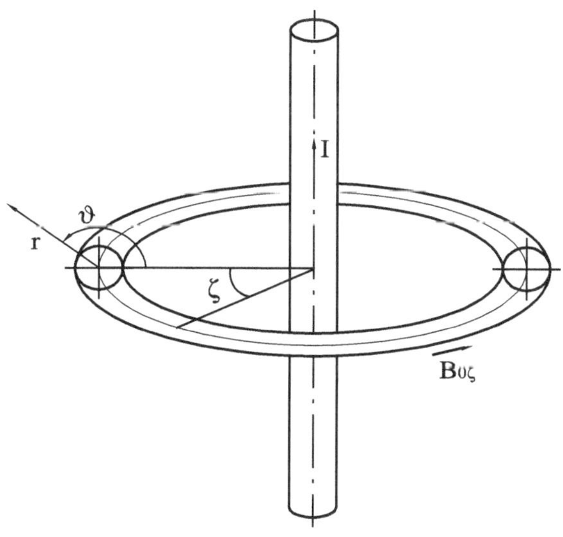

In the framework of magnetic hydrodynamic approximation the vectors of electric induction and electric field are connected by the tensor $\varepsilon_{ij}$ of the dielectric permittivity of cold magneto-active plasma without taking into account collisions between plasma particles. Under the case of small toroidicity (the parameter of toroidicity $\varepsilon_t = a/R << 1$) one can represent expressions for the components of the plasma permittivity tensor $\varepsilon_{ij}$ as a series over the small parameter of toroidicity:

$$\varepsilon_{1,2} = \varepsilon_{1,2}^{(0)} + \varepsilon_{1,2}^{(1)} \cos\vartheta + \varepsilon_{1,2}^{(2)}. \tag{4.26}$$

The main summands in the series (4.26) do not depend on the applied co-ordinates, so $\varepsilon_{1,2}^{(0)} = \varepsilon_{1,2}(B_{0\zeta} = B_0)$. The summands of the first order of smallness over the minor radius of the torus $r$ are as follows:

$$\varepsilon_1^{(1)} = \sum_\alpha \frac{-2\varepsilon_t \Omega_\alpha^2 \omega_\alpha^{(0)2}}{\left(\omega^2 - \omega_\alpha^{(0)2}\right)^2}, \quad \varepsilon_2^{(1)} = -\varepsilon_t \sum_\alpha \frac{\omega_\alpha^{(0)} \Omega_\alpha^2 \left(\omega^2 + \omega_\alpha^{(0)2}\right)}{\omega\left(\omega^2 - \omega_\alpha^{(0)2}\right)^2}, \tag{4.27}$$

here the sub-script $\alpha$ designates the type of the plasma particles (ions $\alpha = i$ and electrons $\alpha = e$), $\Omega_\alpha$ and $\omega_\alpha$ are the plasma and cyclotron frequencies of the $\alpha$-type particles, respectively. The second order corrections are quadratic with respect to the minor radius $r$ of the torus. In the series (4.26) the summands of the second order of smallness, which are proportional to $\exp(2i\vartheta)$ are not taken into account because they make a contribution to the correction of the SFXM frequency which is larger than the second order of smallness. In zero approximation, the cyclotron frequencies $\omega_\alpha$ in the expressions (4.27) are determined by the strength of the toroidal magnetic field (4.25): $\omega_\alpha^{(0)} = \omega_\alpha(\varepsilon_t = 0)$.

The toroidal component of the SFXM magnetic field can be calculated from the following equation, which one can derive from the set of Maxwell equations:

$$\begin{aligned} &\frac{1}{r}\frac{\partial}{\partial r}\left(\frac{r\varepsilon_1}{\varepsilon_\perp}\frac{\partial H_\zeta}{\partial r}\right) + \frac{i}{r}\frac{\partial}{\partial\vartheta}\left(\frac{\varepsilon_2}{\varepsilon_\perp}\right)\frac{\partial H_\zeta}{\partial r} + \frac{\omega^2}{c^2}H_\zeta + \frac{1}{r^2}\frac{\partial}{\partial\vartheta}\left(\frac{\varepsilon_1}{\varepsilon_2}\frac{\partial H_\zeta}{\partial\vartheta}\right) \\ &\quad - \frac{i}{r}\frac{\partial}{\partial r}\left(\frac{\varepsilon_2}{\varepsilon_\perp}\right)\frac{\partial H_\zeta}{\partial\vartheta} = 0, \end{aligned} \tag{4.28}$$

here $\varepsilon_\perp = \varepsilon_1^2 - \varepsilon_2^2$. In zero approximation, Eq. (4.28) describes the independent propagation of SFXMs with different values of azimuthal mode numbers. But in the case of toroidal non-uniformity of an external stationary magnetic field one can take into consideration satellite harmonics of the SFXM field. Taking into the account symmetry of the problem (see Eqs. (4.26), (4.27)) the solution of Eq. (4.28) can be represented in the form of the following wave packet:

$$H_\zeta = \left[H_\zeta^{(0)}(r) + H_\zeta^{(2)}(r) + H_\zeta^{(+1)}(r)\, e^{i\vartheta} + H_\zeta^{(-1)}(r)\, e^{-i\vartheta}\right] \exp(im\vartheta - i\omega t). \tag{4.29}$$

The components of the SFXM electric fields can be written in a form which is similar to expression (4.29), where the next (with respect to the fundamental harmonic $\propto \exp(im\vartheta)$) two satellite harmonics $\propto \exp[i(m \pm 1)\vartheta]$ are taken into account. The field correction $H_\zeta^{(2)}$, which is determined by the waveguide toroidicity, appears as summand of the second order smallness.

For further analysis, it is necessary to determine the relationship between the poloidal component of the wave electric field $E_\vartheta$ and its magnetic component $H_\zeta$:

$$E_\vartheta^{(0)}(r) = \frac{c}{i\omega\varepsilon_\perp^{(0)}}\left[\varepsilon_1^{(0)}\frac{dH_\zeta^{(0)}}{dr} + \frac{m\varepsilon_2^{(0)}}{r}H_\zeta^{(0)}\right], \tag{4.30}$$

$$E_\vartheta^{(\pm1)}(r) = \frac{c}{i\omega\varepsilon_\perp^{(0)}}\left[\varepsilon_1^{(0)}\frac{\partial H_\zeta^{(\pm1)}}{\partial r} + \frac{m\pm1}{r}\varepsilon_2^{(0)}H_\zeta^{(\pm1)} + \frac{\varepsilon_\perp^{(0)}\varepsilon_1^{(1)} - \varepsilon_\perp^{(1)}\varepsilon_1^{(0)}}{2\varepsilon_\perp^{(0)}}\frac{\partial H_\zeta^{(0)}}{\partial r}\right.$$
$$\left. + \frac{m\left(\varepsilon_\perp^{(0)}\varepsilon_2^{(1)} - \varepsilon_\perp^{(1)}\varepsilon_2^{(0)}\right)}{2r\varepsilon_\perp^{(0)}}H_\zeta^{(0)}\right], \tag{4.31}$$

$$E_\vartheta^{(2)}(r) = \frac{1}{\varepsilon_\perp^{(0)}}\left\{-\frac{\varepsilon_\perp^{(2)}}{2}E_\vartheta^{(0)} - \frac{\varepsilon_\perp^{(1)}}{2}\left(E_\vartheta^{(+1)} + E_\vartheta^{(-1)}\right) + \frac{c\varepsilon_1^{(1)}}{2i\omega}\frac{\partial}{\partial r}\left(H_\zeta^{(+1)} + H_\zeta^{(-1)}\right)\right.$$
$$+ \frac{c\varepsilon_1^{(0)}}{i\omega}\frac{\partial H_\zeta^{(2)}}{\partial r} + \frac{c\varepsilon_1^{(2)}}{2i\omega}\frac{\partial H_\zeta^{(0)}}{\partial r} + \frac{cm\varepsilon_2^{(0)}}{i\omega r}H_\zeta^{(2)} + \frac{cm\varepsilon_2^{(2)}}{2i\omega r}H_\zeta^{(0)}$$
$$\left. + \frac{cm\varepsilon_2^{(1)}}{2i\omega r}\left[(m+1)H_\zeta^{(+1)} + (m-1)H_\zeta^{(-1)}\right]\right\}. \tag{4.32}$$

Here

$$\varepsilon_\perp^{(0)} = \varepsilon_1^{(0)2} - \varepsilon_2^{(0)2},\ \varepsilon_\perp^{(1)} = 2\varepsilon_1^{(0)}\varepsilon_1^{(1)} - 2\varepsilon_2^{(0)}\varepsilon_2^{(1)},$$
$$\varepsilon_\perp^{(2)} = \varepsilon_1^{(1)2} + 2\varepsilon_1^{(0)}\varepsilon_1^{(2)} - \varepsilon_2^{(1)2} - 2\varepsilon_2^{(0)}\varepsilon_2^{(2)}. \tag{4.33}$$

To derive the dispersion relation for SFXMs propagating in a toroidal chamber, which is entirely filled with magneto-active plasma, one can apply the standard boundary conditions (see Sect. 2.1 in details).

In zero approximation one can apply the theory of SFXMs propagating along the azimuthal direction in a cylindrical metal waveguide with radius $a$, which is entirely filled with uniform plasma (see Sect. 2.2). The amplitudes of the satellite harmonics can be found as solutions of the considered problem in first approximation ($H_\zeta^{(\pm1)}(r)$ are assumed to be parameters of the first order of smallness). They can be substituted into Eq. (4.28) taking into account the expressions (4.26). Then summands of the first order of smallness will appear there with a factor $\propto \exp[i(m \pm 1)\vartheta]$. That is why accounting for summands of the first order of smallness does not change the amplitudes of the fundamental harmonic and consequently the solution of the considered problem in first order approximation does not determine the SFXM frequency correction. These summands form inhomogeneous Bessel equations for the satellite harmonics $H_\zeta^{(\pm j)}(r)$ with the known right hand parts. These equations have the following solutions, which satisfy the boundary condition that the wave fields should have a finite value on the waveguide axis:

$$H_\xi^{(\pm 1)}(r) = -I_{m\pm 1}(k_\perp r)\int_a^r K_{m\pm 1}(k_\perp r)\hat{L}_\mp^{(1)} I_m(k_\perp r)\,rdr + C_\pm I_{m\pm 1}(k_\perp r) + K_{m\pm 1}(k_\perp r)\int_0^r I_{m\pm 1}(k_\perp r)\hat{L}_\mp^{(1)} I_m(k_\perp r)\,rdr, \tag{4.34}$$

where $C_\pm$ are integration constants and the operator $\hat{L}_\pm^{(1)}$ has the following form

$$\begin{aligned}\hat{L}_\pm^{(1)} = \left(2\varepsilon_1^{(0)}\varepsilon_\perp^{(0)}\right)^{-1}&\left\{\frac{\varepsilon_\perp^{(1)}}{r}\frac{\partial}{\partial r}\left(r\varepsilon_1^{(0)}\frac{\partial}{\partial r}\right) + \frac{\varepsilon_\perp^{(0)}}{r}\frac{\partial}{\partial r}\left(r\varepsilon_1^{(1)}\frac{\partial}{\partial r}\right)\right.\\ &-\varepsilon_1^{(0)}\frac{\partial\varepsilon_\perp^{(1)}}{\partial r}\frac{\partial}{\partial r} - \frac{\varepsilon_\perp^{(0)}}{r}\frac{\partial\varepsilon_2^{(1)}}{\partial r}\hat{\Theta} + \frac{\varepsilon_2^{(0)}}{r}\frac{\partial\varepsilon_\perp^{(1)}}{\partial r}\hat{\Theta} - \frac{\varepsilon_\perp^{(1)}}{r^2}\varepsilon_1^{(0)}\hat{\Theta}^2\\ &-\frac{\varepsilon_\perp^{(0)}\varepsilon_1^{(1)}}{r^2}\hat{\Theta}^2 + 2\frac{\omega^2}{c^2}\varepsilon_\perp^{(0)}\varepsilon_\perp^{(1)} \mp \frac{\varepsilon_\perp^{(0)}\varepsilon_\perp^{(1)}}{r^2}\hat{\Theta} \pm \frac{\varepsilon_1^{(0)}\varepsilon_\perp^{(1)}}{r^2}\hat{\Theta}\\ &\left.\pm\frac{\varepsilon_2^{(1)}\varepsilon_\perp^{(0)}}{r}\frac{\partial}{\partial r} \mp \frac{\varepsilon_2^{(0)}\varepsilon_\perp^{(1)}}{r}\frac{\partial}{\partial r}\right\}, \quad \hat{\Theta}I_m = -mI_m.\end{aligned} \tag{4.35}$$

The integration constants $C_\pm$ can be found from the boundary condition that the amplitudes of the first satellite harmonics of the poloidal electric field vanish at the metal surface of the waveguide, $E_\vartheta^{(\pm 1)}(a) = 0$:

$$\begin{aligned}C_\pm = -&\left[\varepsilon_1^{(0)}k_\perp K'_{m\pm 1}\int_0^a I_{m\pm 1}\hat{L}_\mp^{(1)} I_m rdr + \frac{m\pm 1}{a}\varepsilon_2^{(0)}K_{m\pm 1}\right.\\ &\int_0^a I_{m\pm 1}\hat{L}_\mp^{(1)} I_m rdr + \frac{\varepsilon_\perp^{(0)}\varepsilon_1^{(1)} - \varepsilon_\perp^{(1)}\varepsilon_1^{(0)}}{2\varepsilon_\perp^{(0)}}k_\perp I'_m\\ &\left.+\frac{m\left(\varepsilon_\perp^{(0)}\varepsilon_2^{(1)} - \varepsilon_\perp^{(1)}\varepsilon_2^{(0)}\right)}{2a\varepsilon_\perp^{(0)}}I_m\right]\cdot\left[\varepsilon_1^{(0)}k_\perp I'_{m\pm 1} + \frac{m\pm 1}{a}\varepsilon_2^{(0)}I_{m\pm 1}\right]^{-1}_{\substack{r=a\\ \omega=\omega_m^{(0)}}}.\end{aligned} \tag{4.36}$$

The correction of the SFXM frequency caused by the toroidicity of the applied stationary magnetic field can be calculated in second order approximation of the considered problem. The summands of second order smallness in Eq. (4.29) are proportional to one of the following phase's multipliers: either $\exp(im\vartheta)$ or

$\exp[i(m \pm 2)\vartheta]$. The summands, which have the factor $\exp[i(m \pm 2)\vartheta]$ arrange the equation for the amplitudes of the second satellite harmonics. Thus these summands in series (4.29) can be neglected because their contributions to the SFXM frequency corrections have values of such order of smallness, which are higher than the second order. The summands of the second order of smallness, which are proportional to the factor $\exp(im\vartheta)$ lead to the inhomogeneous differential Bessel equation for the correction $H_\zeta^{(2)}(r)$ to the amplitude of the fundamental harmonic. A solution of this equation, which satisfies the condition of finite value of the wave field on the circular axis of the torus, can be found using the method of constant variation:

$$\begin{aligned} H_\xi^{(2)} = & K_m(k_\perp r)\int_0^r I_m\left(\hat{L}_-^{(1)}H_\xi^{(-1)} + \hat{L}_+^{(1)}H_\xi^{(+1)} + \hat{L}^{(2)}H_\xi^{(0)}\right) rdr \\ & - I_m(k_\perp r)\int_a^r K_m\left(\hat{L}_-^{(1)}H_\xi^{(-1)} + \hat{L}_+^{(1)}H_\xi^{(+1)} + \hat{L}^{(2)}H_\xi^{(0)}\right) rdr + C_2 I_m(k_\perp r). \end{aligned} \tag{4.37}$$

A solution of Eq. (4.37) is the sum of the particular solution of the inhomogeneous differential equation (the first two integral terms) and the general solution of the corresponding homogeneous differential equation (the third term $C_2 I_m(k_\perp r)$). In the general expression (4.29) for the axial component of the wave magnetic field, the term $C_2 I_m(k_\perp r)$ and the amplitude $H_\zeta^{(0)}(r)$ are similar terms. Therefore the constant of integration $C_2$ plays the role of the correction to the normalized multiplier, which in zero order approximation was calculated to be one. That is why the constant $C_2$ cannot be found from the above mentioned boundary conditions, but with the aid of the following condition, which is similar to the condition of normalizing the wave-function in quantum mechanics [26]:

$$\int_0^a \left[2H_\zeta^{(2)} I_m(k_\perp r) + \left(H_\zeta^{(+1)}\right)^2 + \left(H_\zeta^{(-1)}\right)^2\right] rdr = 0. \tag{4.38}$$

Equation (4.38) expresses the following physical condition for the value of the wave field energy: the value of the magnetic field energy calculated taking into account the summands of second order of smallness coincides with the value, which is calculated in zero order approximation. At the same time it can be underlined that the type of the dispersion equation, which is derived by taking into account the summands of second order of smallness does not depend on the value of the integration constant $C_2$.

Application of the boundary condition on the metal wall of the considered toroidal waveguide for the fundamental harmonic of the poloidal component of these wave electric fields and taking into account the summands of second order of smallness leads to obtaining the dispersion equation in the form $D^{(0)} + D^{(2)} = 0$, where $D^{(0)}$ describes SFXMs in the case of uniform stationary magnetic field (see Sect. 2.2) and the second term describes the correction $D^{(2)} \propto E_\vartheta^{(2)}(a)$, which is determined by the toroidicity of the utilized magnetic field $\vec{B}_{0\zeta}$. That is why the correction to the eigen frequency $\omega_m^{(0)}$ of these SFWs can be calculated by expanding

$D^{(0)}$ in Taylor series. So the value of this correction shows to be of the second order of smallness: $\omega_m^{(2)} \propto \varepsilon_t^2$. The condition of applicability of the applied method for determination of the dispersion properties of SFXMs, which propagate in a toroidal metal waveguides, is given by the validity of the inequality $\left|\omega_m^{(2)}\right| \ll \omega_m^{(0)}$.

Therefore the basic mathematical achievement of this result can be formulated in the following way: the problem of two-dimensional non-uniformity of an external stationary magnetic field over the waveguide poloidal cross-section has been reduced to a one-dimensional non-uniformity (this means that finally we have only a radially non-uniform distribution of the considered external magnetic field). But of course, the obtained solutions have a complicated form.

In the limiting case of wide waveguides $R \gg a \gg |m|\, k_\perp^{-1}$, the frequency correction of SFXMs with an eigen frequency expressed by Eq. (2.14) can be represented in the LF range by the following asymptotic expression [27]:

$$\omega_m^{(2)} = -0.25\varepsilon_t^2 \omega_m^{(0)}. \tag{4.39}$$

In the HF range the eigen frequency of SFXMs is given by Eq. (2.15) and their frequency correction can be written with the same accuracy in the following way:

$$\omega_m^{(2)} = \frac{\varepsilon_t^2}{4} \frac{c^2 m^2}{\omega_m^{(0)2} a^2} \left(1 + \frac{3\,|\omega_e|\, k_\perp a}{\omega_m^{(0)}\, |m|}\right) \omega_m^{(0)}. \tag{4.40}$$

For SFXMs which propagate in narrow $|m| \gg ak_\perp$ toroidal waveguides in the LF range one can derive the following asymptotic expression for the frequency correction:

$$\omega_m^{(2)} = \frac{2m^2 - m + 1}{4m^2(m+1)} \varepsilon_t^2 \omega_m^{(0)}. \tag{4.41}$$

In this limiting case the frequency correction for HF SFXMs is approximately equal to:

$$\omega_m^{(2)} = \frac{2m^2 + m + 1}{2m^2} \varepsilon_t^2 \left(\frac{m\delta}{a}\right)^4 \frac{\Omega_e}{|\omega_e|} \omega_m^{(0)}. \tag{4.42}$$

It should be mentioned that the spatial distributions of the satellite harmonics in these two limiting cases are substantially different, for instance in the limiting case of a narrow waveguide: $H_\zeta^{(\pm 1)}(r) \propto r^{m \pm 1}$. However, their values are approximately equal to each other on the metal waveguide wall.

To carry out a comparative analysis of the obtained results let us formulate here asymptotical expressions for the correction $\omega_{el}^{(2)}$ to the eigen frequency of SFXMs which is due to an ellipticity of the transverse cross-section of the metal waveguide, completely filled with plasma. For SFXMs, which propagate in a wide waveguides in the LF range the frequency correction $\omega_{el}^{(2)}$ is as follows:

$$\omega_{el}^{(2)} = -0.25\varepsilon_{el}^2 (4 - m^2 + 2a\delta^{-1})\omega_m^{(0)}. \tag{4.43}$$

In this case the shape of the cross-section of the chamber is determined by the ellipticity parameter $\varepsilon_{el}$ in the following way: $r(\vartheta) = a(1 + \varepsilon_{el} \cos(2\vartheta - \pi))$. In the limiting case of a narrow waveguide, the frequency correction is given by:

$$\omega_{el}^{(2)} = -\frac{\varepsilon_{el}^2}{2m} \frac{2 - 4m^2 - 4m^2}{2 - 4m^2} \frac{a^2}{\delta^2} \omega_m^{(0)}. \tag{4.44}$$

Thus after comparing expressions (4.41) and (4.42) with expressions (4.43) and (4.44), respectively one can make the following conclusions: first, the influence of a toroidal non-uniformity of the external magnetic field on the spectra of transversal SWs is stronger than the influence of an ellipticity of the waveguide cross-section in the case of a rarefied plasma density; second, comparison of the obtained results on the influence of the toroidal change of the magnetic surfaces is redundant, because in experimental practice this change is too small nearby the surface of the metal chamber, where the considered waves are localized [27].

Consequently, the toroidal non-uniformity of the external stationary magnetic field leads to the fact that SFXMs propagate in toroidal waveguides in the form of wave packet. The amplitudes $A_{m+N}$ of the satellite harmonics in such packets are sufficiently small compared with the amplitude of the fundamental harmonic $A_m$ so that $A_{m+N} \sim \varepsilon_t^N A_m$. The influence of the toroidal non-uniformity of the external stationary magnetic field on the amplitude of the fundamental harmonic shows up in the second order of smallness over the parameter of toroidicity $\varepsilon_t$, it can be stronger than the effect of non-circularity of the fusion chambers cross-sections under the determined conditions. The presented results can be useful for practical application in the branch of plasma electronics, and also for interpretation of experimental results concerning phenomena in the peripheral plasma confined by an external magnetic field in fusion devices.

## 4.4 Propagation of Surface Flute Waves Around a Metal Ring in Non-Uniform Toroidal Magnetic Field

In this section within the framework of the method of successive approximation, extraordinarily polarized electromagnetic SFWs with the field components $E_r$, $E_\vartheta$, $H_\zeta$ propagating across a constant toroidal magnetic field around a metal ring are studied. The scheme of the waveguide structure is shown in Fig. 4.5. The minor radius of the metal ring with ideal conductivity is assumed to be small compared with the major radius of the ring, so that the inequality $\varepsilon_t = a/R \ll 1$ is valid. The constant toroidal magnetic field $\vec{B}_0 = B_{0\zeta}\vec{e}_\zeta$ is created by a conductor with a direct current, which is located at the center-line of the ring, perpendicularly to the plane of symmetry of the ring, and the explicit expression for $B_{0\zeta}$ is given by formula (4.25). The plasma, which surrounds this metal ring is assumed to be cold and uniform [28] and all other suppositions, which have been done in Sect. 4.3 are valid in this section as well.

The components of the dielectric permittivity tensor $\varepsilon_{ik}$ can be written in the form of series over the small parameter of toroidicity $\varepsilon_t$ like expression (4.26). In the present case, solution for the toroidal $H_\zeta(r)$ component of the wave magnetic field can be obtained from Eq. (4.28) in the form like expression (4.29). The solution of Eq. (4.28) can satisfy the boundary conditions, which are particularly discussed in Chap. 2. In this case the amplitude $H_\zeta^{(0)}(r)$ of the fundamental harmonic of the considered SFWs in zero order approximation can be expressed by the aid of the MacDonald function [18] $K_m(k_\perp r)$.

Determination of the amplitudes $H_\zeta^{(\pm 1)}(r)$ of the satellite harmonics can be performed by a method, which is similar to the method applied in the previous section, because of the similarity of these problems. Thus solving the present problem in the first order approximation does not change the amplitude of the main mode. Consequently, in the first order approximation, the toroidal variations of the external magnetic field do not contribute to the SFWs eigen frequency. These first-order terms are summands of the inhomogeneous Bessel equations for $H_\zeta^{(\pm 1)}(r)$ with known right-hand side. The noted equations have the following solutions (compare with (4.34)), which satisfy the boundary condition for the SFWs fields at infinite distances from the ring [28]:

$$\begin{aligned} H_\zeta^{(\pm 1)}(r) &= K_{m\pm 1}(k_\perp r)\int\limits_a^r r I_{m\pm 1}(k_\perp r)\hat{L}_\mp^{(1)} K_m(k_\perp r)\,dr \\ &+ C_\pm K_{m\pm 1}(k_\perp r) - I_{m\pm 1}(k_\perp r)\int\limits_\infty^r r K_{m\pm 1}(k_\perp r)\hat{L}_\mp^{(1)} K_m(k_\perp r)\,dr, \end{aligned} \tag{4.45}$$

where the operator $\hat{L}_\pm^{(1)}$ was defined by Eq. (4.35). The integration constants $C_\pm$ can be found from the boundary condition that the amplitudes of the first satellite harmonics of the poloidal electric field vanish on the metal surface of the ring, $E_\vartheta^{(\pm 1)}(a) = 0$ (compare with Eq. (4.36)):

$$\begin{aligned} C_\pm &= \left[\frac{m+1}{a}\varepsilon_2^{(0)} K_{m\pm 1}(k_\perp a) - \varepsilon_1^{(0)} k_\perp K'_{m\pm 1}(k_\perp a)\right]^{-1} \\ &\left\{\varepsilon_1^{(0)} k_\perp I'_{m\pm 1}(k_\perp a)\int\limits_\infty^a r K_{m\pm 1}(k_\perp r)\hat{L}_\mp^{(1)} K_m(k_\perp r)\,dr\right. \\ &+ \frac{m\pm 1}{a}\varepsilon_2^{(0)} I_{m\pm 1}(k_\perp a)\int\limits_\infty^a r K_{m\pm 1}(k_\perp r)\hat{L}_\mp^{(1)} K_m(k_\perp r)\,dr \\ &+ k_\perp K'_m(k_\perp a)\frac{\varepsilon_\perp^{(0)}\varepsilon_1^{(1)} - \varepsilon_\perp^{(1)}\varepsilon_1^{(0)}}{2\varepsilon_\perp^{(0)}} \\ &\left.- \frac{m\left(\varepsilon_\perp^{(0)}\varepsilon_2^{(1)} - \varepsilon_\perp^{(1)}\varepsilon_2^{(0)}\right)}{2a\varepsilon_\perp^{(0)}} K_m(k_\perp a)\right\}\Bigg|_{\omega=\omega_m^{(0)}}. \end{aligned} \tag{4.46}$$

The corrections for the SFXM eigen frequencies in this case can be found by the same method, as it was done in Sect. 4.3. The inhomogeneous differential Bessel equation for the second order correction $H_\zeta^{(2)}(r)$ to the amplitude of the fundamental harmonic of the SFXMs and its solution are similar to the corresponding equation and its solution, which have been obtained in the previous section. This is the reason why one can apply the method of constant variation once again for the present problem (compare with Eq. (4.37)):

$$H_\zeta^{(2)}(r) = K_m(k_\perp r) \int\limits_a^r r I_m(k_\perp r) \left(\hat{L}_-^{(1)} H_\xi^{(-1)} + \hat{L}_+^{(1)} H_\xi^{(+1)} + \hat{L}^{(2)} H_\xi^{(0)}\right) dr$$
$$+ C_2 K_m(k_\perp r) - I_m(k_\perp r) \int\limits_\infty^r r K_m(k_\perp r) \left(\hat{L}_-^{(1)} H_\xi^{(-1)} + \hat{L}_+^{(1)} H_\xi^{(+1)} + \hat{L}^{(2)} H_\xi^{(0)}\right) dr. \tag{4.47}$$

The boundary conditions mentioned above in this section are insufficient to determine the integration constant $C_2$ in the solution (4.47). This constant can be found from a condition similar to (4.38) implying that the energy of the wave magnetic field calculated in the second order approximation should coincide with the energy calculated in the zero order approximation. In quantum mechanics, this requirement is analogous to the normalization condition for the wave function [26].

In the limiting case of thick rings $R \gg a \gg |m|/k_\perp$, the LF SFXM frequency correction $\omega^{(2)}$ coincides with expression (4.39) with the exception of the sign. In the HF range their frequency correction becomes negative, but it coincides with (4.40) in its absolute value. In the waveguide structure of the present geometry (see Fig. 4.5), the studied SFWs can exist in the HF range if the following inequality is valid for the considered plasma parameters $(\delta m/a)^2 < |\omega_e| \Omega_e^{-1}$.

For the eigen frequencies of SFWs, which propagate around a thin $a \ll |m| k_\perp^{-1} \ll R$ metal ring in the LF range, it is possible to derive the following asymptotic expression for the frequency correction:

$$\omega^{(2)} = -\varepsilon_t^2 \left(2m^2 + m + 3\right) \omega_m^{(0)} / 8m^2. \tag{4.48}$$

In this case the SFXM frequency $\omega_m^{(0)}$ is determined by the formula (2.14). In the HF range the value of the SFXM eigen frequency in zero order approximation $\omega_m^{(0)}$ is determined by formula (2.15), and the frequency correction in this case is approximately equal to:

$$\omega^{(2)} = -\varepsilon_t^2 (2m^2 - m + 2)(\delta k_\perp^{-1} a^{-2})^2 m(m-1)\omega_m^{(0)}. \tag{4.49}$$

The radial dependences of the satellite harmonics for these SFWs in the limiting case of thin toroidal metal ring are substantially different from each other: $H_\zeta^{(\pm 1)}(r) \propto r^{-|m \pm 1|}$, but nevertheless these amplitudes appear to be of the same order of smallness on the surface of the metal ring because of the corresponding difference in the numerical factors $C_\pm$.

Consequently, the presence of a toroidal non-uniformity of the external magnetic field results in the fact that SFWs propagate in the form of wave packets. The amplitudes of the satellite harmonics in such packets have small values compared with the amplitude of the fundamental (main) harmonic: $A_{m+N} \sim \varepsilon_t^N A_m$. The influence of the toroidal non-uniformity of the utilized external stationary magnetic field $\vec{B}_0 = B_{0\zeta}\vec{e}_\zeta$ on the amplitude of the fundamental harmonic appears to be of the second order of smallness over the parameter $\varepsilon_t$ of toroidicity. The eigen frequency corrections for these SFWs, which are determined by the toroidal non-uniformity of the $\vec{B}_0 = B_{0\zeta}\vec{e}_\zeta$ field appear to have values of the second order of smallness. They are of the same order of smallness as it was found for the frequency corrections in the previous section, but they have the opposite sign compared with the results obtained there.

Concerning the practical application of the results presented in this section, one can see that such type of a metal ring can be applied as an antenna, which is intended for excitation of transversal SWs, for example with the purpose to sustain a microwave gas discharge.

## 4.5 Propagation of Surface Flute Waves in Weakly Rippled Magnetic Field

An external stationary magnetic field, which confines the plasma in a laboratory device, is often rippled. For instance, magnetic mirror traps have a rippled magnetic field [29], the rippled confining magnetic field in tokomaks is connected with the discreteness of the toroidal magnetic field coils, and a rippled magnetic field is typical for such toroidal systems like the ELMO BUMPY TORUS [30, 31]. It is planned [32] that the so-called "mirror" non-uniformity will be predominant in the confining magnetic field of the modular stellarator reactor HELIAS. If in the expression of the confining magnetic field derived for HELIAS, which is given in [32], anybody replaces the flow co-ordinates by cylindrical co-ordinates then this expression will coincide with the expression of the rippled magnetic field applied in this section. In [33] the eigen electromagnetic perturbations of the thermonuclear plasma of the modular stellarator HELIAS are investigated theoretically.

Thus one can see that such types of problems were actively studied in the past and are under consideration at present time. It should be mentioned for instance, that the theory of propagation, conversion and absorption of magneto-hydrodynamic waves in a plasma column, which is located in a weakly rippled magnetic field has been developed in [34–37]. The waves in a rippled magnetic field propagate in the form of wave packets, which in addition to a fundamental harmonic contain an infinite number of satellite spatial harmonics.

Let us consider propagation of eigen electromagnetic flute waves nearby the border of the plasma column separated by a vacuum layer from circular metal chamber cylinder with ideal conductivity (geometry of the system is particularly described in the Sect. 4.3 and co-ordinate system is shown in Fig. 4.5) [38, 39].

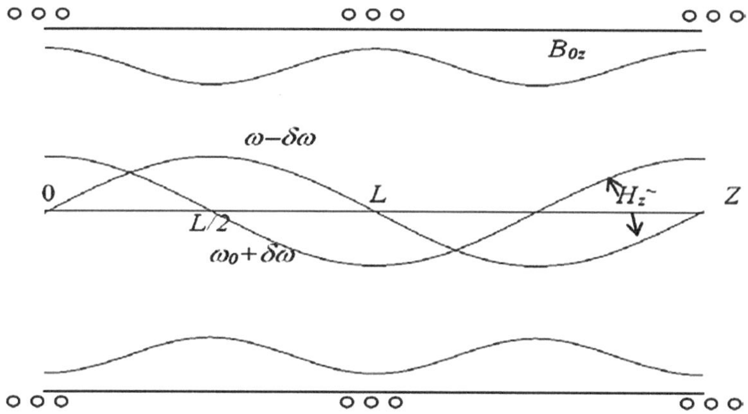


**Fig. 4.6** Scheme of the absolute value of an external stationary magnetic field and standing eigen waves versus the axial coordinate in the case of a weakly rippled magnetic field

The external confining magnetic field $\vec{B}_0 = B_{0z}\vec{e}_z + B_{0r}\vec{e}_r$ is assumed to be weakly rippled (see Fig. 4.6):

$$B_{0r} = B_{00}\varepsilon_m' k_m^{-1} \sin(k_m z), \quad B_{0z} = B_{00}[1 + \varepsilon_m(r) \cos(k_m z)], \tag{4.50}$$

here $\varepsilon_m' \equiv d\varepsilon_m/dr$, $k_m = 2\pi/L$, $L$ is the period of ripples. The characteristic parameter of the ripples $\varepsilon_m$ is usually (for example, in all modern fusion reactors) small. Let us mention for instance, that it is about 0.05 nearby the boundary of plasma confined in the tokomak ASDEX-Upgrade in Garching, Germany [38]; in the HELIAS configuration [33] the "mirror" non-uniformity is planned to be small as well: $\varepsilon_m \sim 0.13$.

One can see that the expressions (4.50) do not automatically provide the implementation of the fundamental equation for fusion plasmas, namely $div\,\vec{B}_0 = 0$. Therefore, one can use expressions (4.50) together with this fundamental equation and then obtain an equation like a modified Bessel equation, which determines the dependence $\varepsilon_m$ on the radial co-ordinate. Its solution is proportional to the modified Bessel function [18] of zero order. Consequently, if the period of the ripples is larger compared to the minor radius of chamber of the fusion device then $\varepsilon_m$ can be considered as a practically constant value. In the opposite case, if the ripple period is small, the $\varepsilon_m$ value decreases approximately exponentially with increasing distance from the plasma interface going to the plasma centre. Thus the non-uniformity of the confining magnetic field is substantial only in a narrow layer within $k_m^{-1}$ thickness nearby the metal wall of the fusion chamber.

As it is known from the basics of electromagnetic theory, the magnetic field fluxes are parallel to the vector of the confining magnetic field induction in every point of the co-ordinate space. The vector form of this condition allows one to derive the next equation for the flux lines in cylinder co-ordinates:

$$dr/B_{0r} = dz/B_{0z}. \tag{4.51}$$

Integrating Eq. (4.51) after inserting the explicit expressions (4.50) for the confining magnetic field components, one can obtain the equation for the magnetic surface in a more convenient way. Taking into account that the properties of the ripple parameter $\varepsilon_m(r)$, which is described by modified Bessel functions one can derive an expression for the radius of magnetic surface:

$$r_0 \approx r + \frac{\cos(k_m z)}{r} \int_0^r r\varepsilon_m dr = r - \cos(k_m z)\left(1 - \frac{\varepsilon_m}{2}\cos(k_m z)\right)\varepsilon\prime_m k_m^{-2} + \mathrm{O}\left(\varepsilon_m^3\right). \tag{4.52}$$

In this case the equilibrium plasma density is a function of the magnetic surface, so one can derive the plasma density as function of the radius of this surface: $n(z,r) = n(r_0)$.

To simplify the consideration of the problem it is suitable to choose convenient a co-ordinate system. Let us apply the system of co-ordinates $\vec{e}_1,\vec{e}_2,\vec{e}_3$ which is related to the flux lines of the magnetic field induction $\vec{B}_0$, namely the first vector is perpendicular to the magnetic surface in every point of the co-ordinate space $\vec{e}_1 = \nabla_{r_0}/\left|\nabla_{r_0}\right|$, the second vector coincides with the azimuthal basic vector in cylinder co-ordinates, and the third vector is parallel to the magnetic field flux lines $\vec{e}_3 = \vec{B}_0/\left|\vec{B}_0\right|$. In this system of co-ordinates the components of the electric field induction $\vec{D}$ and the electric field strength $\vec{E}$ of the wave are coupled by the permittivity tensor of a magneto-active cold plasma without collisions in the simplest way:

$$\vec{D} = \varepsilon_1\left(E_1\vec{e}_1 + E_2\vec{e}_2\right) + \varepsilon_3 E_3\vec{e}_3 - i\varepsilon_2\vec{e}_3 \times \vec{E}. \tag{4.53}$$

Here the components of the cold magneto-active plasma permittivity tensor $\varepsilon_j$ have a standard form [10, 14–16]. The values of the cyclotron frequencies, which are present in expressions for components of the $\varepsilon_j$ tensor, are determined by the complete magnetic field $B_0(r,z)$ and the plasma frequencies, which are applied there as well, can be calculated on the magnetic surface at the radial co-ordinate $r_0$.

Using the symmetry of the problem, the components of the plasma permittivity tensor can be written as series over the small parameter $\varepsilon_m$ of the magnetic field ripples. Let us write down those of them, which describe the extraordinary flute modes taking into account the summands of second order of smallness over the characteristic ripple parameter:

$$\varepsilon_{1,2} = \varepsilon_{1,2}^{(0)}(r) + \varepsilon_{1,2}^{(1)}\cos(k_m r) + \varepsilon_{1,2}^{(2)}(r) + O(\varepsilon_m^3), \tag{4.54}$$

where $\varepsilon_{1,2}^{(0)}(r)$ are the basic summands, and $\varepsilon_{1,2}^{(1)}\cos(k_m r)$ are corrections of the first order of smallness $\left|\varepsilon_{1,2}^{(1)}\right| \sim \left|\varepsilon_m\,\varepsilon_{1,2}^{(0)}\right|$. In second order approximation over the small parameter of the problem one can take into account only corrections to the basic summands $|\varepsilon_{1,2}^{(2)}| \sim |\varepsilon_m^2\,\varepsilon_{1,2}^{(0)}|$, because just they make contributions to the dispersion equation [39] of these SFWs. As a result of calculating the first order and second order approximations of the components of plasma permittivity tensor $\varepsilon_j$ in fluid approximation one can obtain:

$$\varepsilon_1^{(1)} = \frac{\partial \ \ln\left(\Omega_e^2\right)}{\partial r_0}[\varepsilon_1^{(0)}(r) - 1]\frac{\varepsilon_m'}{k_m^2} - 2\varepsilon_m \frac{\Omega_e^2(r)\omega_e^{(0)2}}{(\omega_e^{(0)2} - \omega^2)^2}, \tag{4.55}$$

$$\begin{aligned} \varepsilon_1^{(2)}(r) = &- \frac{\varepsilon_m' \varepsilon_m}{k_m^2} \frac{\partial \ \ln(\Omega_e^2)}{\partial \ r_0}[\frac{\Omega_e^2(r)\omega_e^{(0)2}}{(\omega_e^{(0)2} - \omega^2)^2} + \frac{\Omega_e^2(r)}{4(\omega_e^{(0)2} - \omega^2)}] \\ &+ \frac{\varepsilon_m^2}{2} \frac{\Omega_e^2(r)(3\omega_e^{(0)2} + \omega^2)}{(\omega_e^{(0)2} - \omega^2)^3} - \frac{\varepsilon_m'^2}{2k_m^2} \frac{\Omega_e^2(r)\omega_e^{(0)2}}{(\omega_e^{(0)2} - \omega^2)^2} \\ &+ \frac{\varepsilon_m'^2}{4k_m^2} \frac{\partial^2 \Omega_e^2}{\partial r_0^2} \frac{1}{\omega_e^{(0)2} - \omega^2}, \end{aligned} \tag{4.56}$$

$$\varepsilon_2^{(1)}(r) = \frac{\varepsilon_m'}{k_m^2} \frac{\partial \ \ln \Omega_e^2}{\partial \ r_0} \varepsilon_2^{(0)} + \varepsilon_m \frac{\omega_e^{(0)}\Omega_e^2(r)(\omega^2 + \omega_e^{(0)2})}{\omega(\omega^2 - \omega_e^{(0)2})^2}, \tag{4.57}$$

$$\begin{aligned} \varepsilon_2^{(2)}(r) = &\frac{\varepsilon_m' \varepsilon_m}{2k_m^2} \frac{\omega_e^{(0)}\Omega_e^2(r)\left(\omega^2 + \omega_e^{(0)2}\right)}{\omega\left(\omega^2 - \omega_e^{(0)2}\right)} \frac{\partial \ \ln \Omega_e^2}{\partial r} + \frac{\varepsilon_2^{(0)}}{4\Omega_e^2(r)}\left[\frac{\partial^2 \Omega_e^2}{\partial r^2} \frac{\varepsilon_m'^2}{k_m^4} - \frac{\partial \Omega_e^2}{\partial r} \frac{\varepsilon_m \varepsilon_m'}{k_m^2}\right] \\ &+ \frac{\varepsilon_m'^2}{4k_m^2} \frac{\omega_e^{(0)}\Omega_e^2(r)\left(\omega^2 + \omega_e^{(0)2}\right)}{\omega\left(\omega^2 - \omega_e^{(0)2}\right)^2} + \frac{\varepsilon_m^2}{2} \frac{\omega_e^{(0)3}\Omega_e^2(r)\left(3\omega^2 + \omega_e^{(0)2}\right)}{\omega\left(\omega^2 - \omega_e^{(0)2}\right)^3}. \end{aligned} \tag{4.58}$$

Applying the symmetry of the problem, one can investigate the solution of the Maxwell equations for the SFXM field as a wave packet, which contains the two nearest satellite harmonics proportional to $\exp[i(m\vartheta \pm k_m z)]$ in addition to the fundamental (main) harmonic proportional to $\exp(im\vartheta)$. In the vacuum layer, which separates the plasma column from the metal wall of the waveguide, the radial dependence of the amplitudes of the fundamental harmonic (if the period of magnetic field ripples is sufficiently large, i.e. if the inequality $k_m c < \omega$ is valid) of the wave is expressed through the Bessel functions of the first kind and the Neumann functions [18]. It should be mentioned that this statement concerns the satellite harmonics as well.

As it is known from basic theory of magneto-active plasmas (see e.g. [10, 14–16]), the properties of ordinary modes do not depend on the utilized stationary magnetic field (from mathematical point of view it is explained by the absence of an external magnetic field in the expression for the component of the permittivity tensor $\varepsilon_3$), that is why we shall consider the influence of ripples of the confining magnetic field only on the properties of SFXMs in a plasma cylinder with a non-uniform radial profile of the plasma density. So in zero approximation, it is assumed that only the $m$-th harmonic of the axial component of the SFW magnetic field $H_z^{(0)}(r)$ differs from zero, the other components of the magnetic field are equal to zero as well as the wave axial electric field: $E_z^{(0)}(r) = 0$, $H_\vartheta^{(0)}(r) = H_r^{(0)}(r) = 0$. The dispersion equation for the surface flute X-modes can be derived with the aid of application of the boundary conditions considered in details above in Sect. 2.1; doing that, it is supposed that the waves eigen frequency value is known in zero order approximation.

The inclusion of summands of the first order of smallness in the solution of the Maxwell equations in the plasma region does not change the amplitudes of the fundamental harmonics of the SFXM fields' components, but it results in the appearance of the satellite harmonics, which are proportional to $\exp[i(m\vartheta \pm k_m z - \omega t)]$. Weak ripples of the external confining magnetic field are the reason for the coupling between amplitudes of $E_z^{(\pm)}$satellite harmonics of the axial electric field and amplitudes of the fundamental harmonic of radial electric $E_r^{(0)}$ and axial magnetic $H_z^{(0)}$ fields of the surface flute X-modes. The differential equation, which links these field components of SFXMs, has the form of a inhomogeneous Bessel equation. This is the reason why it is convenient to search for its solution by the method of constant variation. The appearance of the axial electric field of the SFWs, which is not equal to zero is connected also with its weak coupling to the radial electric field because of application of the boundary condition for the tangential electric field on the metal wall of the torus chamber. The amplitudes of the satellite harmonics of the axial magnetic, and also the radial and poloidal electric fields of surface flute X-modes are symmetrical: $E_r^{(+)} = E_r^{(-)}$, $E_\vartheta^{(+)} = E_\vartheta^{(-)}$, $H_z^{(+)} = H_z^{(-)}$ and the amplitudes of the satellite harmonics of the axial electric and also the radial and poloidal magnetic fields are anti-symmetrical: $H_r^{(+)} = -H_r^{(-)}$, $H_\vartheta^{(+)} = -H_\vartheta^{(-)}$, $E_z^{(+)} = E_z^{(-)}$. There is no need to further refining of the spatial distribution of the SFW axial electric field, as well as of the radial and poloidal magnetic fields, because this will merely lead to third- and higher-order corrections to the wave eigen frequency.

By including the second-order terms, one can find the second-order corrections to the amplitudes of the fundamental harmonics of the axial magnetic field of the flute wave and of its radial and poloidal electric fields. The solution of the Maxwell equations for the axial component of the wave magnetic field can be expressed in the following form:

$$H_z^{\sim} = \exp[i(m\vartheta - \omega t)][H_z^{\sim}(r) + H_z^{(+)}(r)\exp(ik_m z) + H_z^{(-)}(r)\exp(-ik_m z)], \tag{4.59}$$

here $H_z^{\sim}(r) = H_z^{(0)}(r) + H_z^{(2)}(r)$, $\left|H_z^{(\pm)}\right| \sim \left|\varepsilon_m H_z^{(0)}\right|$, $\left|H_z^{(2)}\right| \sim \left|\varepsilon_m^2 H_z^{(0)}\right|$. It should be underlined that in formula (4.59), one can ignore the second-order terms which are proportional to $\exp(\pm 2ik_m z)$, because their contribution to the eigen frequency is of a value, whose order is higher than $\varepsilon_m^2$.

The second order correction $H_z^{(2)}(r)$ to the amplitude of the fundamental axial harmonic can be obtained by solving the Maxwell equations for the second order of smallness over the small ripple parameter $\varepsilon_m$:

$$H_z^{(2)} = \psi_0[A_2^{(2)} - \int_a^r \frac{\tilde{\psi}_0 N_X^{(0)2}\hat{Q}H_z^{(0)}dr}{W\left(\psi_0,\tilde{\psi}_0\right)}] + \tilde{\psi}_0 \int_0^r \frac{\psi_0 N_X^{(0)2}\hat{Q}H_z^{(0)}dr}{W\left(\psi_0,\tilde{\psi}_0\right)}. \tag{4.60}$$

Here the square of the transverse refractive index for SFXMs is equal to $N_X^{(0)2} = (\mu^{(0)2} - 1)\varepsilon_1^{(0)}$, $\mu^{(0)} = \varepsilon_2^{(0)}/\varepsilon_1^{(0)}$, $W(\psi_0, \tilde{\psi}_0)$ is the Wronskian of the functions $\psi_0(r)$ and $\tilde{\psi}_0$. The radial profile of the wave magnetic field in zero order approximation, $H_z^{(0)}(r) = A^{(0)}\psi_0(r)$ (herein $A^{(0)}$ being a normalizing factor), is assumed to be known from theory of SFXMs presented in the Chap. 2. It satisfies the boundary condition, that the SFW magnetic field is of finite value inside the waveguide and, in particular on the axis of plasma column $r = 0$. The function $\tilde{\psi}_0$ is a solution of the same equation like the $\psi_0(r)$ function, but it is linearly independent from $\psi_0(r)$ and has a singularity on the column axis. The operator $\hat{Q}$ in solution (4.60) has the following form:

$$\hat{Q}H_z^{(0)} = \frac{N_\vartheta}{N_X^{(0)2}}\left(-q_1 + i\mu^{(0)}q_2\right) + \frac{c}{\omega}\frac{d}{r\,dr}\left[\frac{r}{N_X^{(0)2}}\left(\mu^{(0)}q_1 - iq_2\right)\right], \tag{4.61}$$

here $N_\vartheta = cm/(r\omega)$ is the poloidal refractive index,

$$\begin{aligned} q_1 = &- \varepsilon_m N_\vartheta H_z^{(+)} + \varepsilon_m N_m H_\vartheta^{(+)} + \frac{c\varepsilon_m'}{\omega\, rk_m}\left[\frac{d}{dr}\left(rH_\vartheta^{(+)}\right) - imH_r^{(+)}\right] - \varepsilon_m\varepsilon_1^{(0)}E_r^{(+)} \\ &- \varepsilon_1^{(1)}E_r^{(+)} - \varepsilon_m\varepsilon_1^{(1)}E_r^{(0)}/2 - \varepsilon_1^{(2)}E_r^{(0)} + i\varepsilon_m'\varepsilon_1^{(0)}E_z^{(+)}/k_m - i\varepsilon_2^{(1)}E_\vartheta^{(+)} - i\varepsilon_2^{(2)}E_\vartheta^{(0)} \\ &- i\varepsilon_m\varepsilon_2^{(1)}E_\vartheta^{(0)}/2 - i\varepsilon_m\varepsilon_2^{(0)}E_\vartheta^{(+)} - i(\varepsilon_m')^2\varepsilon_2^{(0)}E_\vartheta^{(0)}/(4k_m^2), \end{aligned} \tag{4.62}$$

$$\begin{aligned} q_2 = &- \varepsilon_1^{(1)}E_\vartheta^{(+)} - \varepsilon_1^{(2)}E_\vartheta^{(0)} + i\varepsilon_2^{(0)}E_r^{(0)}\left[\frac{\varepsilon_m^2}{2} - \frac{(\varepsilon_m')^2}{4k_m^2}\right] + i\varepsilon_2^{(1)}E_r^{(+)} \\ &+ i\varepsilon_2^{(2)}E_r^{(0)} + \varepsilon_2^{(0)}\frac{\varepsilon_m'}{k_m}E_z^{(+)}. \end{aligned} \tag{4.63}$$

Note that the constant of integration $A_2^{(2)}$ in solution (4.60) is a small value of the second order of smallness. If we insert the correction given by formula (4.60) into expression (4.59) describing a wave packet, we shall see that the term $A_2^{(2)}\,\psi_0(r)$ is similar to $H_z^{(0)}$. That is why the constant $A_2^{(2)}$should be determined not from the boundary conditions but rather from a condition analogous to the normalization condition for the wave function in perturbation theory in quantum mechanics [26]. In this case, the indicated condition implies the equality of the electromagnetic wave energy calculated in zero order approximation and in the second one. On the other hand, the form of second order dispersion relation and the correction to the SFXM eigen frequency are independent on the constant $A_2^{(2)}$.

The inclusion of summands of the second order of smallness into the consideration enables one to determine the values of the corrections of the second order of smallness to the amplitude of the fundamental harmonic of the axial magnetic field and also the radial and poloidal electric fields of the SFWs. Thus the spatial distribution of the SFXM fields can be determined with the aid of the method of successive approximations from the spatial distribution of the axial magnetic field of the SFWs which is known in the zero approximation within accuracy up to summands of the second order of smallness inclusively. Applying all the above mentioned boundary conditions one can derive the dispersion equation for the surface flute X-modes as it has been done in the previous sections.

It can be analytically studied only in some limiting cases, for example, in the case of a uniform dense plasma ($\Omega_e^2 \gg \omega_e^2$) with a large radius ($a \gg |m|\delta$, where $\delta$ is the skin-depth), which is separated from the metal wall of the torus chamber by a narrow vacuum layer ($R_2 R_1^{-1} - 1 \ll 1$). If this plasma filled waveguide is immersed into an external stationary magnetic field with a large ripple period ($k_m R_1 \ll 1$), the SFXM eigen frequency can be determined in the following way: $\omega = \omega^{(0)} + \Delta\omega$, where $\omega^{(0)}$ is the value of the SFXM frequency in zero order approximation calculated with an accuracy up to summands of the order of $\Delta^2 \ll 1$:

$$\omega^{(0)} \approx \frac{(1-\Delta)\delta^2 m|\omega_e|}{2a^2\Delta - a^2\Delta^2 + 2a\delta} + \frac{\sqrt{(1-2\Delta)\delta^4 m^2\omega_e^2 - (2a^2\Delta - a^2\Delta^2 + 2a\delta)m^2c^2\Delta(3\Delta-2)}}{2a^2\Delta - a^2\Delta^2 + 2a\delta}, \tag{4.64}$$

and $\Delta\omega$ is a correction to this value, which is connected with the weak ripples of the confining magnetic field of the fusion device:

$$\begin{aligned}\frac{\Delta\omega}{\omega^{(0)}} \approx -\frac{(\varepsilon_m')^2}{4k_m^4}\Bigg[&\frac{k^2\delta}{a}\Big(1 - \frac{m\delta|\omega_e|}{a\omega^{(0)}}\Big)\Big(\frac{m^2}{a^2k^2} - 1\Big) \\ &+ \frac{\Delta}{\delta^2} - \frac{\Delta m^2\delta^{-2}}{a^2k^2} - \frac{3m^2}{k^2a^4} + \frac{\delta + a}{a^2\delta} - \frac{m|\omega_e|}{a^2\omega^{(0)}}\Bigg] \\ &\times \left[(1-\Delta)\frac{m\delta^2|\omega_e|}{a^2\omega^{(0)}} + (2-3\Delta)\frac{\Delta m^2}{k^2a^2}\right]^{-1}.\end{aligned} \tag{4.65}$$

Summarizing the material represented in this section one can emphasize that a similar mathematical approach has been applied for investigation of beam-plasma instabilities, which occur in waveguides with a periodic dielectric inserts and/or in waveguides with corrugated walls [40, 41]. In those cases the dispersion equations, which describe the excitation of such waveguides by charged particle beams, contain small correcting summands, which are proportional to the square of the ripple depth of the device. A similar correction, but proportional to the squared amplitude of an external static electric field that varies periodically along the axis of a magnetized waveguide, enters into the dispersion relation describing the emission of electromagnetic wave power from an electrostatic free-electron laser. The fact that the same mathematical methods for solving problems related to the different branches of the electrodynamics of dense electron beams, as well as of the electrodynamics of plasmas in modular stellarators, yield dispersion relations with similar structures provides one more independent confirmation of the validity and correctness of the results obtained in the present sections devoted to propagation of SFXMs in waveguides immersed into spatially non-uniform magnetic fields of different geometry. Experimental study of the processes of modes conversion in a waveguide with corrugated walls proves that for practical aims it is sufficient to apply analytical calculations, which have been done within the framework of the method of successive approximations [40].

## 4.6 Conclusions

The results of theoretical studies of the properties of SFXMs under the conditions of their propagation in plasma filled cylindrical non-uniform waveguides of different constructional peculiarities are presented here. Real experimental conditions, namely spatial non-uniformity of the confined plasma density and of the utilized external magnetic field are taken into account. Possible practical applications of these theoretical results are indicated here as well.

The performed investigations allow us to determine the frequencies of the eigen electromagnetic flute waves and spatial distribution of their field analytically, calculate expressions for their damping rates, which are connected both with collisions between the plasma particles and resonant conversion of these modes into bulk modes in the case of a dense plasma. It has been proved that if the plasma density increases in the direction towards the plasma border, the LF SFXM frequency value decreases as compared with the case of a uniform plasma density profile. It has been shown that approximation of the periphery plasma density profile by a linear function (independently on its form in the central region of the plasma column) allows one to find the LF SFXM frequency spectrum analytically. In the case of a linear type of the periphery plasma density profile the obtained expression for their frequency as function of the plasma parameters is similar to that one, which was obtained in the case of a uniform plasma when doing a replacement of the SFXM penetration depth by a definite effective value.

It has been found that radial non-uniformity of an external axial magnetic field has an influence, which is similar (from mathematical point of view) to the case of the influence of a plasma density radial non-uniformity, on the frequency of these modes. A small toroidicity of an external magnetic field (parameter $\varepsilon_t \ll 1$) results in the feature that SFWs propagate as wave packets. Corrections to the wave field of their fundamental harmonic and to their eigen frequency, which are determined by this non-uniformity of the utilized magnetic field, are proportional to $\varepsilon_t^2$.

Ripples of the confining magnetic field in modern thermonuclear reactors and in special devices for plasma electronics are characterized by a small parameter $\varepsilon_m \ll 1$. It is shown that such ripples can influence conversion and absorption of the eigen flute waves. In this case SFWs propagate as well as wave packets, in which all six field components are present. Nevertheless this problem can be solved by the aid of the method of successive approximation. It has been proved that this problem is analogous (from mathematical point of view) to the problems of beam instabilities in a plasma, which is confined in a waveguide with a periodic dielectric insert or with corrugated walls and even to the theory of free electron lasers. The dispersion relations obtained for the SFXMs in this case contain small summands, which are proportional to $\varepsilon_m^2$. They can be adequately solved by the methods of successive approximations.

## References

1. Anderson, T. (2011). *Plasma antennas*. Boston: Artech House.
2. Jenn, D. C. (2003). *Plasma Antennas: Survey of Techniques and the Current State of the Art.* SPAWAR PMW 189, San Diego, CA.
3. Kumar, V., Mishra, M., & Joshi, N. K. (2011). Study of a fluorescent tube as plasma antenna. *Progress in Electromagnetics Research Letters, 24*, 17–26.
4. Kudrin, A. V., Petrov, E Yu., Kyriacou, G. A., et al. (2005). Insulated cylindrical antenna in a cold magneto-plasma. *Progress in Electromagnetics Research, 53*, 135–166.
5. Wu, M., Wen, B. Y., Zhou, H. (2009). Ionospheric clutter suppression in HF surface wave radar. *Journal of Electromagnetic Waves and Applications*, **23**(10), 1265–1272.
6. Aliev, Yu M, Schluter, H., & Shivarova, A. (2000). *Guided-wave-produced plasmas*. New York: Springer.
7. Margot, J., Moisan, M. (1992). *Surface wave sustained plasmas in static magnetic fields for study of ECR discharge mechanisms—Microwave Excited Plasmas*. In: M. Moisan, J. Pelletier (Ed.) Elsevier, Amsterdam.
8. Winter, R., Korzec, D., Engemann, J. (1997) Large area plasma cleaning with 26" microwave slot antenna plasma source SLAN II. Surface and Coating Technol., **93**(1), 134–141.
9. Korzec, D., Werner, F., Winter, R., et al. (1996). Scaling of microwave slot antenna (SLAN): A concept for efficient plasma generation. *Plasma Sources Science and Technology, 5*, 216–234.
10. Alexandrov, A. F., Bogdankevich, L. S., & Rukhadze, A. A. (1984). *Principles of plasma electrodynamics*. Berlin: Springer.
11. Vukovic, V. (1986). *Surface waves in plasmas and solids*. Singapore: World Scientific.
12. Beletski, N. N., Bulgakov, A. A., Khankina, S. I., & Yakovenko, V. M. (1984). *Plasma instabilities and non-linear phenomena in semi-conductors*. Kyiv: Naukova dumka. (in Russian).
13. Agranvich, V. M., & Mills, D. L. (1982). *Surface Polaritons*. Amsterdam: North-Holland.
14. Akhiezer, A. I., Akhiezer, I. A., Polovin, R. V., Sitenko, A. G., & Stepanov, K. N. (1975). *Plasma electrodynamics*. Oxford: Pergamon Press.
15. Krall, N. A., & Trivelpiece, A. W. (1973). *Principles of plasma physics*. New York: MacGraw-Hill.
16. Brambilla, M. (1998). *Kinetic theory of plasma waves. homogeneous plasma*. Oxford: Clarendon Press.
17. Girka, V., Girka, I. (1990). Influence of plasma inhomogeneity on the spectra of azimuthal surface waves. *Radiophysics and Quantum Electronics*. **33**(4), 516.
18. Abramowitz, M., Stegun I. A. (Ed.) (1964). *Handbook of Mathematical Functions*. National Bureau of Standards, Applied Mathematics, Series 55, New York.
19. Girka, V., Girka, I. (1991). Azimuthal surface waves in a non-uniform plasma cylinder. *Radiophysics and Quantum Electronics*. **34**(4), 386.
20. Girka, V., Girka, I., Kondratenko, A., et al. (1988). Azimuthal surface waves of magneto-active plasma waveguides. *Journal of Communication Technology and Electronics,* **33**(8), 37–41.
21. Kondratenko, A. M., & Kuklin, V. M. (1988). *Basis of plasma electronics*. Moscow: Energoatomizdat. (in Russian).
22. Azarenkov, N. A., & Ostrikov, K. N. (1991). Surface magnetoplasma waves at the interface between a plasma-like medium and a metal in a Voigt geometry. *Physics Reports, 308*, 333–428.
23. Beidler, C. D., Grieger, G., Harmeyer, E., et al. (1995). *Helias reactor studies*. Preprint of Max-Planck-Institut für Plasmaphysik, 85748 Garching bei München, FRG, IPP 2/330.
24. Girka, V. (2006). Theory of transverse surface electro-magnetic waves propagating along a plasma–metal boundary with a finite radius of curvature in a magnetic field. *Plasma Physics Reports*, **32**(5), 401–410.

25. Girka, V., Girka, I., Pavlenko, I. (1997). Surface waves propagating in the direction transverse to the axis of magnetized plasma—filled waveguides with noncircular transverse cross-sections. *Plasma Physics Reports,* **23**(11), 959–963.
26. Landau, L. D., Lifshits, E. M. (1977). *Course of Theoretical Physics. Quantum Mechanics. Vol. 3.* Oxford: Pergamon Press.
27. Girka V., Girka I., Girka A., et al. (2011). Theory of azimuthal surface waves propagating in non-uniform waveguides. *Journal of Plasma Physics*, **77**(4), 493–519.
28. Girka, I. (2002). Azimuthal surface waves at the plasma-metal boundary in a non-uniform toroidal magnetic field. *Technical Physics*, **47**(7), 845–850.
29. Rjutov, D. D. (1988). Opened traps. *Uspekhi fizicheskih nauk,* **4**, 565–614 (in Russian).
30. Baity, F. W., Berry, L. A., Bighel, L., et al. (1983). *Plasma properties and ion heating in EBT-S and hot electron rings at TRW.* Proceedings of 9-th Conference on Plasma Physics and Controlled Nuclear Fusion Research, Baltimore, **2,** 185.
31. Fujiwara, M., Kamimura, R., Hosokawa, M., et al. (1983). Experimental and numerical studies on plasma confinement in Nagoya Bumpy Torus (NBT). *Ibid,* 197.
32. Beidler, C. D., Kolesnichenko, Y. I., Marchenko, V. S., et al. (2001). Stochastic diffusion of energetic ions in optimized stellarators. *Physics of Plasmas*, **8**, 2731.
33. Kolesnichenko, Y. I., Lutsenko, V. V., Wobig, H., et al. (2001). Alfven continuum and high—frequency eigenmodes in optimized stellarators. *Physics of Plasmas*. **8**, 491.
34. Girka, I., Stepanov, K. (1992). MHD plasma oscillations in a bumpy magnetic field. *Ukrainian Physical Journal*, **37**(1), 69–75 (in Russian).
35. Girka, I., Lapshyn, V., Stepanov, K.(1998). Splitting of the spectra of MHD plasma oscillations in a rippled magnetic field. *Plasma Physics Reports*. **24**(11), 948–955.
36. Girka, I., Lapshyn, V., Stepanov, K. (1994). Plasma heating near satellite Alfven resonances in confinement systems with a ripple magnetic field. *Plasma Physics Reports,* **20**(11), 916–922.
37. Beljaev, M., Girka, I., Gritsyna, V. (2003). Effect of the periodic ripple in an axial confining magnetic field on the Alfven heating of a cylindrical plasma. *Plasma Physics Reports,* **29**(5), 399–406.
38. Girka, I. O. (1996). *How ripples of the steady magnetic field in tokamaks influence propagation, conversion and absorption of Alfven and fast magneto-sonic waves.* Preprint of Max-Planck-Institut für Plasmaphysik, 85748 Garching bei München, FRG, IPP 4/274.
39. Girka, V., Girka, I. (2006). Additional ECR heating of a radially inhomogeneous plasma via the absorption of a satellite harmonics of the surface flute modes in a rippled magnetic field. *Plasma Physics Reports*, **32**(12), 1047–1051.
40. Kuzelev, V. M., & Rukhadze, A. A. (1990). *Electrodynamics of dense electron beams in plasma.* Moscow: Nauka. (in Russian).
41. Shestopalov, V. P. (1976). *Diffraction electronics.* Kharkiv: Vyscha Shkola. (in Russian).

# Chapter 5
# Surface Flute Waves Propagating in Waveguides with Non-Circular Cross-Section

The most effective way to excite waveguide structures (including those which are either corrugated or contain phase velocity slowing down elements like combs, disks, etc.) can be realized in the range of their eigen frequencies [1–3]. This fact explains the interest to studying frequency spectra of multi-component plasmas filled waveguides supplied with different type slow-wave structures. Multi-component waveguide structures are widely used [4] in modern devices of plasma electronics. The presence of different semiconductor and/or dielectric insertions in a metal waveguide increases the number of eigen modes and power of electromagnetic radiation that can be obtained from generators built up on the basis of these type of waveguides. It also leads to the appearance of some specific features both in the frequency spectra of the waves propagating in such waveguides and in the spatial distribution of their fields compared with the case of application of uniform waveguides. In particular, different types of surface waves (SWs) can propagate there [5]. Properties of SWs, which propagate along a planar boundary between two different media, and properties of SWs propagating along a boundary, which is characterized by a finite value of curvature radius, differ substantially. For instance, in the case of the planar shape of the plasma boundary most of the SWs are slow waves, they are either electrostatic (i.e., their magnetic field is negligible small compared with electric field) or at least can be considered in the electrostatic approach, see, e.g., [6–9] unlike the case of SWs propagating along the plasma boundary with finite value of curvature radius. The frequency ranges of their existence are different for these two types of plasma interfaces.

The shape of the cross-section of the plasma-vacuum interface in fusion devices also distinguishes from a circular one [10, 11]. For example, the shape of the poloidal cross-section of the boundary of closed magnetic surface in the stellarator Wendelstein 7-X [12], which is under construction in Germany, changes along the axial co-ordinate. It periodically (there are five periods) has quasi-triangular, bean-like and intermediate shapes. Thus, when carrying out research into the propagation of axially non-symmetrical SWs in fusion devices one has to take into account the non-circularity of their chamber cross-section. The authors of the references [13–16]

V. Girka et al., *Surface Flute Waves in Plasmas*, Springer Series on Atomic, Optical, and Plasma Physics 79, DOI: 10.1007/978-3-319-02027-3_5,

have made a conclusion that propagation of different types of SWs can be considered as the main reason of undesirable phenomena observed in fusion devices during various RF heating regimes like: heating of peripheral plasma regions, enhancement of interaction between plasma particles and the inner walls of the chamber, contamination of the plasma by impurity ions and consequent damage of the chamber. Thus, investigation of the properties of transverse SWs propagating in magneto-active plasma filled metal waveguides, which have non-circular cross-section, is important as well for analyzing the experimental data obtained in fusion devices.

Using an expansion into Fourier series one can simulate any real shape of the cross-section, which characterizes any experimental plasma device. If the coefficients in this series are small then one can apply the method of successive approximations for studying the problem of SW propagation in such corrugated waveguides. The theory of surface flute waves (SFWs), which has been studied in Chap. 2 can be applied as zero approximation while carrying out this investigation. Spatially periodical non-uniformity of the waveguides, which are under consideration here, is the reason that in this case, electromagnetic waves propagate in the form of wave packets. The existence of a small parameter of the problem allows one: first, to assume that the considered wave packets can be represented in the form of a superposition of the fundamental harmonic and two nearest satellite harmonics; second, to formulate non-linear boundary conditions for the considered SFW fields in the form of Taylor series over this small parameter. This investigation allows one: to calculate the correction to the SFW frequency, whose value in zero approximation is obtained in Chap. 2 (for the case of circular cross-section) and to analyze the spatial distribution of the SFW fields in the case of longitudinally corrugated waveguides.

Let us start our consideration from the description of the general geometrical features of the waveguides, where the SFW can propagate and from the formulation of boundary conditions, which are applied for solving the problem. In the general case, the radius of the waveguide metal chamber with arbitrary shape of its cross-section can be determined by the following series:

$$R_2 = b \cdot \left[ 1 + \sum_{n=1}^{\infty} h_n \sin\left(n\varphi - \varphi_{2n}\right) \right], \tag{5.1}$$

here $b$ is the mean value of radius of the considered waveguide metal chamber and $h_n \ll 1$ is the relative depth of the longitudinal corrugation of the chamber wall. Expression (5.1) is an example of a Fourier series, which can describe any waveguide cross-section. Such a corrugated waveguide, generally speaking, can be filled with a uniform plasma column with radius $R_1$, whose shape is described by the following expression:

$$R_1 = a \cdot \left[ 1 + \sum_{n=1}^{\infty} \delta_n \cos(n\varphi - \varphi_{1n}) \right], \tag{5.2}$$

here $a$ is the mean radius of the plasma column. The parameters $\delta_n \ll 1$ characterize the maximum dimensionless deviation of the plasma column radius from its mean

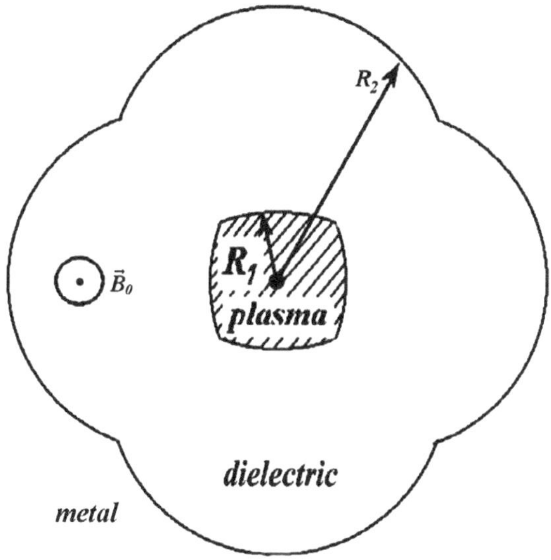


**Fig. 5.1** Scheme of the magneto-active plasma filled waveguide, which has non-circular cross-sections of the chamber and plasma column

value $a$. The corrugation parameters in the expressions (5.1) and (5.2), satisfy the following inequality: $\delta_n, h_n < (b/a - 1) < 1$. The parameters $\varphi_{1n,2n}$ are the phases of the corrugation of the inner surface of the waveguide and the plasma interface, respectively.

In this Chapter the plasma is assumed to be uniform, magneto-active and cold, and that it is separated by a thin ($\Delta = R_2 R_1^{-1} - 1 \ll 1$) dielectric layer with dielectric permittivity constant $\varepsilon_d$ from the wall of the waveguide chamber. The dielectric permittivity of the plasma is described in fluid approximation [1, 17, 18]. An external constant uniform magnetic field is oriented along the axis of the waveguide: $\vec{B}_0 \parallel \vec{z}$. The general geometry of a waveguide with corrugated form of its metal chamber and plasma column, which also has non-circular cross-section, is presented in Fig. 5.1.

To derive the dispersion relation for SFWs propagating in such a waveguide one can apply the following non-linear (over the both parameters $h_n$ and $\delta_n$, which determine the waveguide wall and plasma column interface corrugations) boundary conditions:

$$E_{r,\varphi}(r = 0) < \infty, \quad H_z(r = 0) < \infty, \tag{5.3}$$

(these conditions mean that the SFW fields have finite values in the volume occupied by the plasma column);

$$\begin{gathered} E_\tau(R_1 + 0) = E_\tau(R_1 - 0), \quad E_\tau(R_1) \propto R_1(\varphi) E_\varphi(R_1) + E_r(R_1) \frac{dR_1}{d\varphi}, \\ H_z(R_1 + 0) = H_z(R_1 - 0), \end{gathered} \tag{5.4}$$

(these conditions mean continuity of the wave tangential electric and magnetic fields on the plasma-dielectric interface);

$$E_\tau\left(R_2\right) \propto R_2\left(\varphi\right) E_\varphi\left(R_2\right) + E_r\left(R_2\right)\frac{dR_2}{d\varphi} = 0, \tag{5.5}$$

(this condition means that the tangential electric field on the inner metal wall of the waveguide chamber is zero).

For electromagnetic perturbations with zero axial wave number ($\partial/\partial z = ik_3 = 0$), the set of Maxwell equations splits into two subsets, which describe independent propagation of the X-mode with the field components $E_r, E_\varphi, H_z$ and the O-mode with other three field components. However, deviation of the cross-section of the plasma-dielectric interface from circular shape does not lead to coupling between X- and O-modes and surface flute O-modes cannot propagate in waveguides with a thin dielectric layer ($b - a \ll a;\ b - a \ll c/\omega$) as proved in Chap. 2. That is why we restrict our presentation of the material devoted to propagation of eigen waves in waveguides with non-circular cross-section only to the case of flute X-mode propagation.

## 5.1 Propagation of Surface Flute Waves in Corrugated Metal Waveguides Entirely Filled with Plasma

Let us start our consideration from the case of SFW propagation in longitudinally corrugated metal waveguides which are entirely filled with dense ($|\omega_e| \ll \Omega_e$) uniform cold magneto-active plasma. So in this particular case, the line $R_1(\varphi)$ coincides with the $R_2(\varphi)$ one in Fig. 5.1 and there is no region of dielectric.

At first, one can study the limiting case, when the shape of the waveguide cross-section is described by (5.1), where only one term $n = N$ is taken into account from the complete summation over the index $n$. Then the obtained results will be generalized for the case of a waveguide with arbitrary cross-section. The components of the electric field of the considered SFW can be expressed using the set of Maxwell equations through the axial component of the wave magnetic field in the following manner:

$$E_r = \frac{\omega}{ck_\perp^2}\left(\mu\frac{\partial H_z}{\partial r} + \frac{m}{r}H_z\right), \quad E_\varphi = \frac{i\omega}{ck_\perp^2}\left(\frac{\partial H_z}{\partial r} + \mu\frac{m}{r}H_z\right), \tag{5.6}$$

here $k_\perp^2 = k^2(\mu^2 - 1)\varepsilon_1$, and $\mu = \varepsilon_2/\varepsilon_1$, where $\varepsilon_j$ are components of tensor of dielectric permittivity of the cold magneto-active plasma [1, 17, 18].

In first order approximation with respect to $h_N$, the solution of the set of Maxwell equations for the waves' axial component of the magnetic field has the following form:

$$\begin{aligned} H_z(r,\varphi,t) = \Big[&I_m\left(k_\perp r\right) + C_+^{(1)} I_{m+N}\left(k_\perp r\right)\exp\left(iN\varphi\right) \\ &+ C_-^{(1)} I_{m-N}\left(k_\perp r\right)\exp\left(-iN\varphi\right)\Big]\exp\left(i\left(m\varphi - \omega t\right)\right), \end{aligned} \tag{5.7}$$

here $C_{\pm}^{(1)}$ are constants of integration. From the physical point of view they are amplitudes of the SFW satellite harmonics.

After using expression (5.7) for the calculation of $E_r$ and $E_\varphi$ in Eq. (5.6) these fields can be inserted in boundary condition (5.5). This allows one to forecast that in linear approximation (with respect to $h_N$) the SFW eigen frequency does not change in comparison with the case of waveguides with circular cross-section. But the fields of the SFW even in linear approximation can be expressed in the form of wave packets, which contain the satellite harmonics. Their amplitudes $C_{\pm}^{(1)}$ have the following forms:

$$C_{\pm}^{(1)} = \frac{-h_N}{2D_{m\mp N}^{(0)}} I_m(k_\perp b)\left[k_\perp^2 b^2 + m(m \pm N)\left(1-\mu^2\right)\right]_{|\omega^{(0)}}. \quad (5.8)$$

Thus SFWs propagate in the form of wave packets, which contain spatial harmonics with specified values of the azimuthal mode numbers $m = M \pm jN$, where $j$ are natural numbers. These wave packets always have those satellites, which propagate in the direction, which is opposite to the direction of propagation of the fundamental harmonics. As one can see from formula (5.7) such a satellite harmonic appears already in the first approximation over $h_N$, if $|M| < N$. In the opposite case, if $|M| > N$, such a satellite harmonic appears in one of the higher order approximations. The part of the SFW energy, which is transferred to these satellites, has a value of the order $h_N^{2k}$ (as compared with the main mode energy), where $k = 1$ for the case $|M| < N$ and $k = 2$ for the case $N < M < 2N$ and so on. Therefore, SFW properties are different from those of SWs, which propagate in corrugated waveguides of planar geometry [7, 19].

In second order approximation with respect to the small parameter of the problem $h_N$ the boundary condition for the tangential component of the SFM electric field (5.5) allows one to derive the following equation:

$$\begin{aligned}&(1+h_N \sin N\varphi)E_\varphi(b) + h_N N\cos N\varphi \cdot E_r(b) + b(h_N \sin N\varphi + h_N^2 \sin^2 N\varphi)\\&\times\frac{dE_\varphi}{db} + h_N^2 b \sin N\varphi \cos N\varphi \cdot \frac{dE_r}{db} + \frac{b^2}{2}h_N^2 \sin^2 N\varphi \frac{d^2E_\varphi}{db^2} = 0.\end{aligned} \quad (5.9)$$

Summands, which are proportional to $\exp(\pm 2Ni\varphi)$, can be neglected in Eq. (5.9) because of their smallness. Then one can derive a set of three equations for calculation of the factors, which are in front of the exponents for the different arguments. Its solution can be written in the following form:

$$D_0 + D_{mn}^{(2)} = 0, \quad (5.10)$$

where $D_0(\omega_0) = 0$ is the dispersion equation of SFWs propagating in metal waveguide with circular cross-section, which is entirely filled with plasma (see Sect. 2.2), $D_{mn}^{(2)}$ is a correction to $D_0$, which is connected with spatial azimuthal non-uniformity of the waveguide interface. The superscript (2) means that it is proportional to the square of the small parameter of the problem. In the present case $D_{mn}^{(2)}$ has the following form:

$$
\begin{aligned}
D_{mn}^{(2)} &= h_n^2 I_m (k_\perp b) \left(m^2 + 3k_\perp^2 b^2 - \mu^2 m^2\right)/4 \\
&+ h_n C_+^{(1)} \left[I'_{m+n}(k_\perp b) k_\perp b\, (1+\mu m) + I_{m+n}(k_\perp b)\left(k_\perp^2 b^2 + (m+n)(m+\mu)\right)\right]/2 \\
&+ \frac{h_n}{2} C_-^{(1)} \left[I'_{m-n}(k_\perp b) k_\perp b\, (1+\mu m) + I_{m-n}(k_\perp b)\left(k_\perp^2 b^2 + (m-n)(m+\mu)\right)\right]\Big|_{\omega=\omega_0} .
\end{aligned} \tag{5.11}
$$

An analysis of the dispersion relation (5.10) for a wave packet, whose fundamental harmonic is characterized by the azimuthal mode number $m = M$, proves that in the limiting case of a narrow waveguide, $b \ll M\delta$, the value of the SFW frequency $\omega = \omega_0 + \Delta\omega_N$ becomes less as compared with the case of the circular cross-section, $\Delta\omega_N < 0$. Therefore, spatial azimuthal non-uniformity of the surface of the waveguide wall studied here can be considered as slowing down structure. In the case of wide waveguides $b \gg M\delta$, the longitudinal corrugation of the plasma-metal interface decreases the value of the SFW angular frequency only for SFWs with small values of azimuthal mode numbers $M < N$. If the opposite inequality is valid $M > N$, then in the range $\left(2M/\left(M^2 - N^2\right) < M\delta\, b^{-1} \ll 1\right)$, deviation of the cross-section of the plasma-metal interface from a circular shape leads to increasing of the SFW angular frequency, $\Delta\omega_N > 0$, in comparison with the case of waveguides with circular cross-section.

A numerical analysis proves that the SFW frequency correction is with good accuracy proportional to the external magnetic field, $\Delta\omega_N \sim B_0$, and that the method of successive approximations cannot be applied in the limiting case of very wide waveguides $k_{ef} \to 0$, because of strong increasing of the $\Delta\omega_N$ value.

The investigation performed above allows one to make the conclusion that the value of the SFW frequency correction caused by the deviation of the cross-section of the plasma-metal interface from a circular one does not depend in the second approximation with respect to the small parameter $h_N$ on the presence of other small terms of the Fourier series in the right hand side of expression (5.1). Therefore, in the case of a waveguide with an arbitrary cross-section, the SFW frequency correction caused by the deviation of the cross-section shape of the plasma-metal interface from a circular one can be found as sum: $\Delta\omega = \sum_n \Delta\omega_n$, where the particular corrections $\Delta\omega_n$ can be calculated for different meanings of the index $n$. Every particular frequency correction can be calculated in the following way:

$$
\Delta\omega_n = -D_{mn}^{(2)} \left(\frac{\partial D_0}{\partial\omega}\right)^{-1}_{|\omega=\omega_0}, \tag{5.12}
$$

In limiting cases one can derive specific simple analytical expressions for $\Delta\omega_N$. In the case of wide waveguides the correction of angular frequency can be represented in the following form:

$$
\frac{\Delta\omega_N}{\omega_0} \approx -\frac{h_N^2}{2}\left\{1 + \frac{2N\,|\omega_e|}{\omega_0}\right\}. \tag{5.13}
$$

In the case of narrow waveguides one can obtain another expression for the correction of angular frequency:

$$\frac{\Delta\omega_N}{\omega_0} \approx \frac{h_N^2}{2}\left\{1 - \frac{2N\omega_0}{|\omega_e|} + \frac{4N(\omega_0^2 - \omega_e^2)}{kb\Omega_e}\right\} < 0. \tag{5.14}$$

Analyzing the expressions (5.13) and (5.14) one can compare them with the case of waveguides with circular cross-sections which is considered in Sect. 2.2. As one can see, the increase of the strength of an external magnetic field leads to a decrease of the SFW frequency in comparison with the case considered in Sect. 2.2. This effect of an external magnetic field is more pronounced in the limiting case of a narrow waveguide with a non-circular cross-section than in the case of a wide waveguide.

Since SFWs in longitudinally corrugated waveguides propagate in the form of wave packets the spatial distribution of their fields is also different from the case of their propagation in waveguides with circular cross-section. Results of numerical analysis of the spatial distribution of the SFW magnetic field in such type of waveguides are presented in Fig. 5.2. There one can see the value of the envelope of the SFW magnetic field on the metal wall of the waveguide as function of the azimuthal angle for the following values of the waveguide parameters: $b = 2\delta$, $\Omega_e^2 = 50\omega_e^2$, N $= 4$, $h_4 = 0.05$. Increasing azimuthal mode number $m$ leads to decreasing magnitude of the wave magnetic field. The number of maxima of the SFW magnetic field $H_z$ located along the metal wall of the waveguide is equal to the value of the azimuthal mode number. The numbers of the locations on the spatial distribution of $H_z$, where one can see contributions of satellite harmonics is

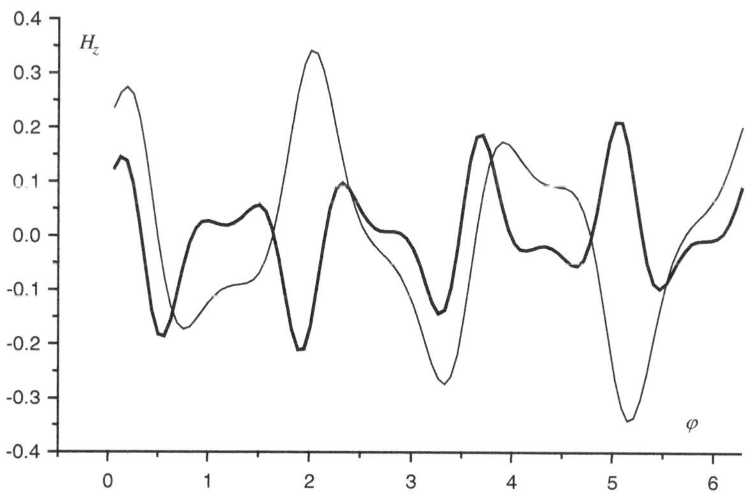


**Fig. 5.2** Angular distributions of the axial magnetic field of SFXMs on the metal wall of a corrugated waveguide for the cases $m = 5$ (*thick line*) and $m = 3$ (*thin line*). In the case $m = 5$ the field amplitude was multiplied by ten for convenience of the reader

equal to value of the corrugation parameter $N$, which determines the periodicity of the corrugation.

From the practical point of view the results described above in this section have one more important application, it can be used for studying SFW propagation across an external axial magnetic field in a waveguide with square cross-section (square waveguide). It is interesting that in this case application of the Cartesian system of co-ordinates seems to be more suitable: first, this is dictated by reasons of symmetry; second, it has allowed one already to solve similar problems in the cases of semi-bounded plasma and planar layer (see e.g., [19]). Nevertheless, application of the Cartesian system of co-ordinates does not allow one to separate variables in the case of a square waveguide, so let us apply cylindrical co-ordinates.

The equation of the surface of a square waveguide with the side $a$ in the cylindrical system of co-ordinates looks like:

$$R_2 = b\,(1 + h_4 \cos(4\varphi) + h_8 \cos(8\varphi) + h_{12} \cos(12\varphi) + \cdots), \tag{5.15}$$

here $b = (a/\pi)\ln\left(3 + 2\sqrt{2}\right)$, $h_4 = -0.1394$, $h_8 = 0.04395$, $h_{12} = -0.02082$. In order to calculate the SFW frequency correction according to expression (5.12) one has to neglect the terms of the third order of smallness. Then the applicability of expression (5.15) in the case of a square waveguide can be estimated by the value of the product $h_4^2 h_8 = 8.54 \times 10^{-4}$.

Let us consider as an example [20] the square waveguide with the side $a = 1.8$ cm, which is filled with *n-Ge* semiconductor (density of conductivity electrons $n_e = 2.5 \times 10^{13}$ cm$^{-3}$, effective mass of electrons $m_e = 1.3 \times 10^{-27}$ g, dielectric permittivity of its lattice is 16. It is immersed into a constant magnetic field $B_0 = 100$ G. In physics of semiconductors such SWs are named as surface magneto-plasma polaritons, and such geometry of waveguide is entitled as Voigt geometry [7, 8, 21]. The SFWs with the azimuthal mode number $m = 1$ have the lowest eigen frequency, its value is: $\omega = \omega_0(m = 1) + \Delta\omega_4 + \Delta\omega_8 + \Delta\omega_{12} + \cdots \approx 2.23 \times 10^9$ s$^{-1}$. The penetration depth of the SFW field into the semiconductor material *n-Ge* is $k_\perp^{-1} = 0.134$ cm, so its value is small compared with the waveguides' transversal sizes. The life-time of conductivity electrons $\tau = 10^{-5}$ s in *n-Ge* is larger than the SFW period, which justifies the application of the collisionless plasma model here.

The presented example proves that for practical needs to calculate the frequency of SFWs, which propagate in a square waveguide, it is sufficient to take into account [in addition to the basic term $\omega_0(m)$] only the first correction, because the value of the terms, which are neglected in this case, is about 4 %. However, such situation is not of universal character. For example, this statement is not valid for waveguides, whose cross-sections have a rectangular shape with the sides $d$ and $b$, respectively.

Let us assume that the difference of these dimensions is relatively small, so $|b - d| \ll b$. Then to estimate the value of terms, which are neglected in this case, one can calculate the asymptotic limit of expression (5.12) for large values of the parameter $n$. That means, if the inequality $n \gg k_\perp b \gg |m|$ is valid, one gets for low frequency ($\omega < |\omega_e|$) SFWs:

$$\Delta\omega_n = -0.5h_n^2 k_\perp b\, n\, \omega_0(m). \tag{5.16}$$

For high frequency SFWs ($\omega_{UH} < \omega < \omega_2$, here $\omega_2 = |\omega_e|/2 + \sqrt{\Omega_e^2 + \omega_e^2/4}$ is the cut-off frequency and $\omega_{UH}$ is the upper hybrid resonance frequency), the frequency correction has another form:

$$\Delta\omega_n \approx -0.5h_n^2 k_\perp b\; n\; \omega_e^2 \;/\; \omega_0(m). \tag{5.17}$$

An analysis of expressions (5.16) and (5.17) shows that the expression for $\Delta\omega_n$ in the form of series over $h_n$ converges, if this small parameter decreases faster than $n^{-1}$ with increasing parameter $n$.

Consequently, the SFW frequency in a square waveguide with the side $a$ is close to the SFW frequency, in a round cylindrical waveguide with radius $R = (a/\pi)\ln\left(3 + 2\sqrt{2}\right)$. For determination of the SFW frequency in zero approximation it is possible to apply the asymptotic formulas for SFWs obtained in Sect. 2.2. Then its approximate value is: $\omega_0(m = 1) = 2.477 \times 10^9\,\text{s}^{-1}$. This value differs from the value of the SFW frequency calculated above less than by 2 %. This is a good result, but the approximate value of the frequency correction calculated by the help of expression (5.17) is $\Delta\omega_4 = -3.7 \times 10^7\,\text{s}^{-1}$ which differs from the value obtained numerically from the SFW dispersion equation by an order of magnitude. This noticeable difference is explained by the fact that $k_\perp b = 7.5$ in this case, so that the inequality, which confirms the validity of the approximate expression (5.17) is only badly fulfilled [20].

Let us point out two differences of the properties of SFWs, which propagate in a square waveguide from those SWs, which propagate in a planar plasma layer. First, the SW frequency in a uniform plasma layer does not depend on the thickness of the waveguide (in hydro-dynamic approach), but the SFW frequency in a square waveguide explicitly depends on the waveguides' sizes. This is explained by the fact that the frequency of SWs propagating in a plasma layer depends on the wave vector along the direction of their propagation. In the case of a plasma layer, which has infinite size along the direction of SW propagation, this vector can be of arbitrary value. But in the case of SFWs propagating in a square waveguide the role of the wave vector is played by the ratio $(M/b)$ for the fundamental harmonic. Second, SWs propagate along the opposite surfaces of the planar plasma layer independently from each other, whereas the SFW phases on the opposite surfaces of a square waveguide correlate strongly (the difference of the phases for the fundamental harmonic is equal to $M\pi$). Thus, one can make the conclusion that due to the finite longitudinal sizes of the uniform plasma layer, which fills a rectangular waveguide, there exists a connection between SFWs, which propagate along its opposite surfaces.

The spatial distributions of the magnetic and radial electric fields of the SFW, whose fundamental harmonic is characterized by the azimuthal mode number $m = 1$, in a waveguide filled with $n - Ge$ are shown in Figs. 5.3 and 5.4. The graphs are calculated according to the formulas (5.6) and (5.7), in which small

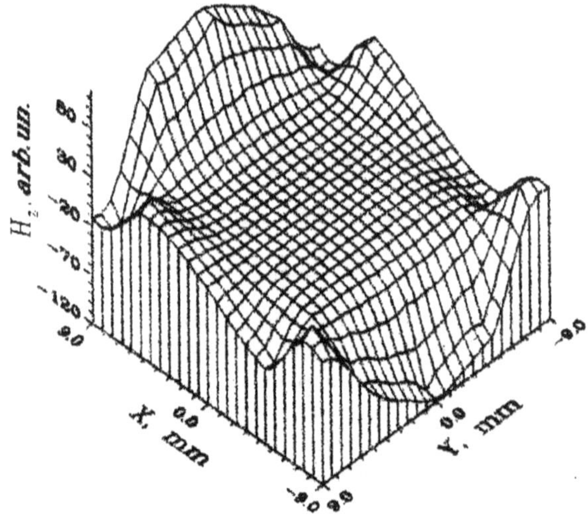

**Fig. 5.3** Spatial distribution of axial magnetic field of a SFW in the cross-section of a square waveguide, which is entirely filled with cold magneto-active plasma

satellites $\propto \exp\left(i\left(m \pm 4\right)\varphi\right)$ and $\propto \exp\left(i\left(m \pm 8\right)\varphi\right)$ are taken into account. The magnitudes of the fields are indicated in arbitrary units. One can see that the SFW magnetic field has maximal amplitude: this means that SFWs are non-electrostatic, or in other words they are electromagnetic waves. Their field consists of a very large magnetic component and a very small electric component, which can be even neglected. The numerical analysis shows that the amplitude of the fast satellite $\propto \exp\left(i\left(M + N\right)\varphi\right)$ is larger than the amplitude of the slow satellite $\propto \exp\left(i\left(M - N\right)\varphi\right)$ if the inequality $M < N$ is valid, and vice versa, if the opposite inequality $M > N$ is fulfilled.

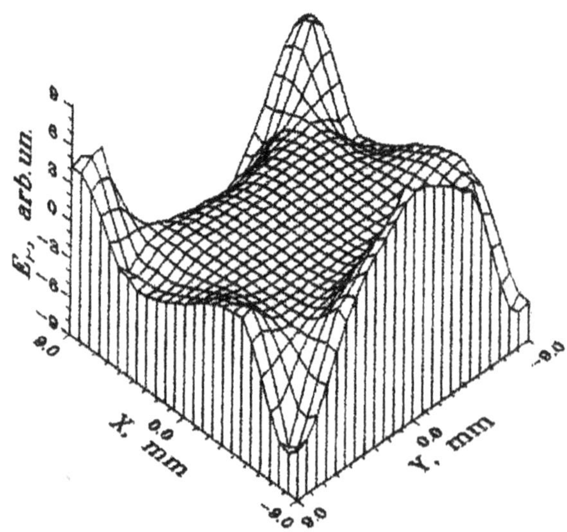

**Fig. 5.4** Spatial distribution of radial electric field of a SFW in the cross-section of the square waveguide which is entirely filled with cold magneto-active plasma

The dependence of the SFW frequency correction in a rectangular waveguide on the density of charged particles, constant magnetic field and linear dimensions of the waveguide is analogous to that one studied for a square waveguide. To demonstrate the applicability of this method let us consider LF SFXMs propagating in a waveguide with cross-section $b \times a = 1 \times 2$ mm, which is filled with *n-Ge* semiconductor with the parameters indicated above. It should be underlined that if one can take into account the first four terms on the right hand side of Eq. (5.15), then it would be possible to describe rectangular waveguides, which sides satisfy the equality $a = 2b$ with very high accuracy. For instance, the area of the waveguide cross-section in this approximation is only by 4 % less than the area of the proper rectangle. In this case the frequency of LF SFXMs with the azimuthal mode number of the fundamental harmonic $m = 1$ can be calculated by the following way: $\omega \approx \omega_0(m = 1) + |\omega_e|\,(\Delta\omega_2 + \Delta\omega_4 + \Delta\omega_6 + \Delta\omega_8)$. So its value is $\omega \approx 1.78 \cdot 10^{10} - 1.95 \times 10^{10} \cdot (0.034 + 1.05 \times 10^{-4} + 2.842 \times 10^{-3} + 2.132 \times 10^{-3})$ $= 1.707 \times 10^{10}\text{s}^{-1}$.

Therefore, the ratio of the frequency correction and the basic term is about 4.3 %. One can see that the period of these LF SFXMs is smaller than the life-time of conductivity electrons $\tau \approx 10^{-5}$ s in a typical $n$-semiconductor (this confirms the validity of the application of the collisionless plasma model for this problem).

The properties of SFWs in such type of waveguide are similar to the properties of SFWs in a waveguide with circular cross-section that has the radius $R \approx 0.0766$ cm. The penetration depth of the SFW field into $n$-Ge is $k_{\perp}^{-1} \approx 0.142$ cm, which is larger than the radius $R$. The difference between the analytical value of the SFW eigen frequency ($1.773 \times 10^{10}$ s$^{-1}$) and the value obtained by numerical calculation of the SFW dispersion relation is less than 1 %. In wide waveguides, where $k_{ef} < 0.3$, the eigen frequency of LF SFXMs becomes small, and the absolute value of its correction increases. Therefore, in this limiting case the proposed method is not applicable.

The numerical analysis of the spatial distribution of the SFW fields shows that the amplitude of their magnetic component in a rectangular waveguide with the parameters considered above is one order larger than the amplitudes of their electric fields. The amplitudes of their satellites $\propto \exp(-i\varphi)$ and $\propto \exp(-i3\varphi)$, which propagate in reverse direction as compared with the fundamental harmonic, relate to the amplitude of the fundamental harmonic as 4 % and 0.1 %, respectively. Propagation of the other group of satellites, which are proportional to $\propto \exp[i(m + 2j)\varphi]$, coincides with the direction of the fundamental harmonic propagation, but their phase velocity is less than the velocity of the fundamental harmonic. The amplitudes of these slow satellites are larger than the amplitudes of the satellites, which propagate in opposite direction. For example, the amplitudes of the satellites $\propto \exp(+i3\varphi)$ and $\propto \exp(+i5\varphi)$ relate to the amplitude of fundamental harmonic as 18 % and 1.2 %, respectively.

Finally, it should be noted that SFWs can propagate as well nearby the surface of a metal rectangular bar, which is oriented parallel to an external magnetic field and coated by thick (as compared with the penetration depth of the SFW field into this solid state plasma) layer of $n$-semiconductor. Consequently, for this case one can also apply the described method and predict that the main results will be analogous to those presented here.

## 5.2 Propagation of Surface Flute Waves in Metal Waveguide with Non-Circular Cross-Section Partially Filled with Plasma

Let us consider a waveguide, whose radius is described by expression (5.1); it is filled with a cold uniform plasma column with radius $R_1 = a$; an external magnetic field is oriented along the $\vec{z}$ axis. Thus, to illustrate the geometry of the waveguide in this particular case one can also apply the scheme presented in Fig. 5.1, but the line $R_1(\varphi)$, which describes the shape of the plasma column cross-section must be changed to be a circle. Solving the set of Maxwell equations for extraordinarily polarized SFWs (SFXMs) in the plasma region, one can find expressions for their fields in the form of series (5.7). The components of their electric field are expressed by the formulas (5.6). In the dielectric region, $H_z(r,t)$ is determined by the superposition of Neumann functions and Bessel functions of the first kind, see expression (2.2) in Sect. 2.1. Remind that in this region of space, the components of the SFXM electric field are expressed through $H_z(r,t)$ by formulas (5.6), where one can make the following changes: $\mu \to 0$, $k_\perp^2 \to -k^2$.

In zero approximation one can apply the theory presented in [22]. Then the SFXM magnetic field can be written in the form of just one definite harmonic and the dispersion relation has the form of the SFXM dispersion relation considered in Chap. 2. In the first approximation one takes into account the terms $\propto h_n$ in the expressions for the SFXM fields. Application of the boundary conditions indicated before at the beginning of Sect. 5.1 allows one to derive the dispersion relation in first order approximation. Its solution for the SFW frequency can be written in the form of a sum like it has been done in the previous section. Using the recurrent relations for cylindrical functions that allows one to rewrite Bessel functions or their derivatives with the aid of the Bessel functions of another order and assuming the dielectric layer to be thin, it is possible to find [23] analytical expressions for $\Delta\omega_N$, which have a simple form in the limiting cases.

In a wide ($k_\perp b, \kappa_d b \gg |m|$) waveguide, the correction of angular frequency is equal to:

$$\frac{\Delta\omega_N}{\omega_0} \approx -\frac{h_N^2}{16}\left\{1 - \frac{2k_\perp^2 a^2 (k_\perp a + \varepsilon_2 m \varepsilon_1^{-1})}{(k_\perp a + \varepsilon_2 m \varepsilon_1^{-1})^2 - \varepsilon_2^2 N^2 \varepsilon_1^{-2}}\right\}, \tag{5.18}$$

and in a narrow waveguide ($k_\perp b, \kappa_d b \ll 1$), the correction of angular frequency is equal to:

$$\frac{\Delta\omega_N}{\omega_0} = \frac{h_N^2}{16}\left\{1 - \frac{2\,(\varepsilon_2 - \varepsilon_1)\, m^3}{\varepsilon_d\left(m^2 - N^2\right)}\right\}. \tag{5.19}$$

Analyzing the expressions (5.13) and (5.14) one can compare them with the case of waveguides with circular cross-sections which is considered in Sect. 2.2. As one can see, the increase of the strength of an external magnetic field leads to a decrease of the SFW frequency in comparison with the case considered in Sect. 2.2. This effect of an external magnetic field is more pronounced in the limiting case of a narrow waveguide with a non-circular cross-section than in the case of a wide waveguide.

## 5.3 Influence of Plasma Column Cross-Section Non-Circularity on Dispersion Properties of Low Frequency Surface Flute Waves Propagating in Magneto-Active Waveguides

Let us consider a cylindrical metal chamber with radius $b$ and thin dielectric coating with dielectric permittivity $\varepsilon_d$ on its inner surface. An external magnetic field is oriented along the z-axis: $\vec{B}_0 \parallel \vec{z}$. The considered waveguide is uniform along its axis and is filled with a cold plasma column with sufficient density ($\Omega_e \gg |\omega_e|$). The shape of the plasma cross-section is assumed to be arbitrary [see expression (5.2)]. So to illustrate the geometry of the plasma filled waveguide in this particular case one can apply Fig. 5.1 once again, but now the line $R_2(\varphi)$, which describes inner shape of the metal chamber, must be changed to be a circle.

In zero approximation, the plasma-dielectric interface is considered as circular and concentric with the metal chamber, so one can apply the theory of SFWs presented in Chap. 2. But periodic spatial non-uniformity of the plasma surface (5.2) leads to propagation of the SFWs in the form of wave packets which propagate not only along the azimuthal direction but along radial direction as well. Such a packet contains (in addition to the fundamental harmonic $\propto \exp(im\varphi)$) an infinite number of satellite harmonics $\propto \exp[i(m \pm jn)\varphi]$, $j = 1, 2, 3 \ldots$. According to the theory of electromagnetic waves propagating in waveguides with one-dimensional periodicity [24] the amplitudes of satellite harmonics are lower than the amplitude of the fundamental harmonic, their ratio is $\propto \delta_n^j$. To solve the problem of LF SFW propagation in the considered waveguide in second order approximation over the small parameter $\delta_n$ one can consider firstly a simpler shape of the plasma-dielectric interface, $R_1 = a[1 + \delta_N \cos(N\varphi + \varphi_N)]$, and then generalize the obtained results for the case of arbitrary shape of the plasma-dielectric interface (5.2). In this approximation it is sufficient to take into account in the wave packet only the fundamental harmonic and its two nearest satellites.

The radial dependence of the SFW harmonics in a dielectric region is expressed through Bessel functions of the first kind and Neumann functions. In the plasma region, they are determined by modified Bessel functions and their derivatives over the argument [25]. Within the accuracy of second order of smallness over the small parameter of the problem, the boundary conditions on the surface $r = a$ can be written in the following form:

$$[E_\varphi(1 - h_N^2 N^2) - E_r h_N N \sin(N\varphi + \varphi_N) + a h_N \cos(N\varphi + \varphi_N)\frac{\partial E_\varphi}{\partial r} + \frac{a^2}{2} h_N^2 \cos^2(N\varphi + \varphi_N)\frac{\partial^2 E_\varphi}{\partial r^2}]|_a = 0 \quad (5.20)$$

$$[H_z + a h_N \cos(N\varphi + \varphi_N)\frac{\partial H_z}{\partial r} + \frac{a^2}{2} h_N^2 \cos^2(N\varphi + \varphi_N)\frac{\partial^2 H_z}{\partial r^2}]\Big|_a = 0. \quad (5.21)$$

The dispersion relation in zero approximation: $D_0(\omega_0) = 0$, can be derived by application of the boundary conditions on the circular plasma surface:

$$D_0 = I_m(k_\perp a)\beta_2 - \alpha_2(r = a)\beta_1, \quad (5.22)$$

where

$$\alpha_2(r) = J_m'(\kappa_d b)N_m(\kappa_d r) - J_m(\kappa_d r)N_m'(\kappa_d b); \quad \kappa_d = k\sqrt{\varepsilon_d}; \quad k_\perp = k\sqrt{(\mu^2 - 1)\varepsilon_1};$$

$$\beta_1 = m\mu I_m(k_\perp a)/(ka\varepsilon_d) + I_m'(k_\perp a)\sqrt{\mu^2 - 1}/\varepsilon_d; \quad \beta_2 = -\alpha_2'(r = a)/\sqrt{\varepsilon_d}.$$

Regarding only terms of the first order of smallness in the boundary conditions (5.20) and (5.21) allows one to determine the amplitudes of SFW satellite harmonics. An analysis proves that they have no symmetry: $H_z^{(+)} \neq \pm H_z^{(-)}$. Taking into consideration the terms $\propto \exp(im\varphi)$ in Eqs. (5.20) and (5.21), one can derive the SFW dispersion relation within an accuracy of the second order smallness over $h_n$: $D_0 + D_n^{(2)} = 0$, here $D_n^{(2)}$ is a second order correction. The explicit form of the $D_n^{(2)}$ is very cumbersome (see Ref. [26] in details) so we shall not show it here; instead of that the same equation, but for the case of HF flute X-modes is derived in Sect. 5.5.

The LF SFW dispersion relation can be solved analytically only in some limiting cases. For example, in the case of wide plasma waveguides ($a/\delta \gg |m|, N$), their angular frequency is determined as a sum $\omega = \omega_0 + \Delta\omega_N$, where

$$\omega_0 \approx \frac{\sqrt{(1 - 2\Delta)\delta^4 m^2 \omega_e^2 - (2a^2\Delta - a^2\Delta^2 + 2a\delta)m^2 c^2 \Delta(3\Delta - 2)}}{(2a^2\Delta - a^2\Delta^2 + 2a\delta)} + \frac{\sqrt{\varepsilon_d} + (1 - \Delta)\delta^2 m|\omega_e|}{(2a^2\Delta - a^2\Delta^2 + 2a\delta)} \quad (5.23)$$

$$\Delta\omega_N \approx -h_N^2 \omega_0 \frac{3m^2\kappa_d^2 a^2 + 2\kappa_d^4 a^4 - 3m^4 + m^2 N^2}{4m^2(N^2 - 4m^2)\Delta^2}. \quad (5.24)$$

The expression (5.24) contains the resonant denominator $N^2 - 4m^2$, thus it cannot be applied in the case $N = 2|m|$. This is associated with the fact that the spectra of SFWs propagating in waveguides with circular cross-section without external magnetic field are degenerated with respect to the sign of the azimuthal mode number. Analysis of this ($N = 2|m|$) resonant case requires additional studies; its results are presented in the next Sect. 5.4.

One can see from (5.24) that the frequency correction increases with decreasing dielectric layer thickness. However, this increase is limited by the inequality $\Delta > h_n$. Thus, the SFW frequency can be represented as a series over the small parameter $\delta_n$, if $\delta_n$ decreases faster than $n^{-3/2}$ with increasing summarizing index $n$ [26].

The results of our numerical studies are shown in the Figs. 5.5 and 5.6. They prove that for wide waveguides ($k_{ef} \ll 1$) dispersion curves $\Delta\omega_n(k_{ef})$ weakly depend on $m$. Unlike in the case of narrow waveguides there exists a non-monotonous character of the curves. On these plots one can see that in the region of small effective wave numbers $k_{ef} \leq 0.5$, there is no essential difference between the curves $\Delta\omega_1(k_{ef})$ calculated for positive and negative values of the mode number $m$. This numerical analysis confirms the conclusions of the analytical investigations that the $\Delta\omega_n$ value increases with decreasing thickness and dielectric permittivity of the dielectric layer and increasing $B_0$. The values of $\Delta\omega_n$ are different for SFWs propagating in mutually opposite directions. It should be noted that in the considered case, the frequency of SFWs with $m > 0$ is higher than the frequency of SFWs with $m < 0$, the other parameters being the same [27]. This difference is small for

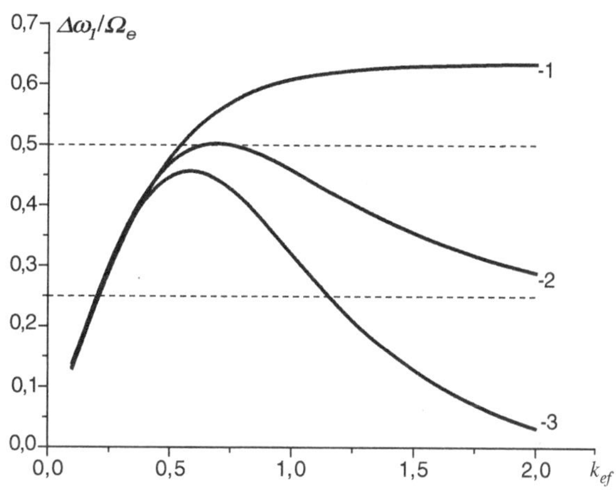


**Fig. 5.5** Corrections to the LF SFW eigen frequency in the non-resonant case for the azimuthal mode numbers $m = -1, -2, -3$; N = 1

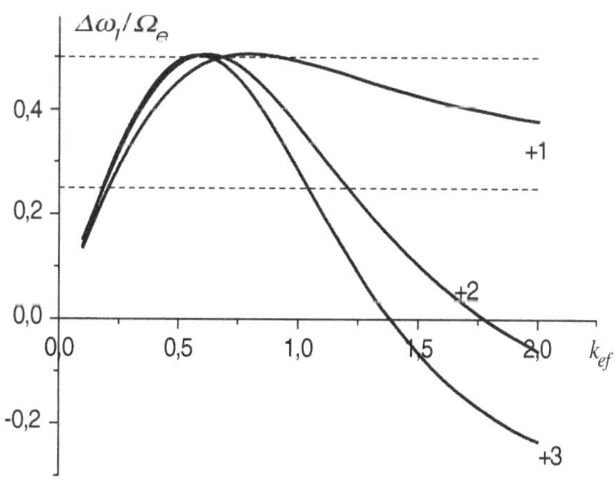


**Fig. 5.6** The same as in Fig. 5.5, but for positive values of $m$

wide waveguides, but it increases with increasing value of $|m|$ and for the case of waveguides with a narrow dielectric layer.

A study of the spectral composition of SFW wave packets allows one to conclude the following. First, the amplitudes of fast SFW satellite harmonics are lower than amplitudes of their slow harmonics. Second, the amplitudes of the satellite harmonics are larger in wide waveguides. Third, the magnitudes of satellite harmonics decrease with decreasing $|m|$. But in the case of narrow waveguides for $N = 1$ (case of eccentric plasma waveguide) for SFWs with $m = -1$, the amplitude of the standing wave satellite harmonic, which does not transfer its energy along the azimuthal angle, is larger than the amplitude of the slow satellite harmonic, which propagates along the azimuthal angle with the phase velocity $\omega/(m - N) = -\omega/2$ which is less than the velocity of the fundamental harmonic [26]. In general, the effect of corrugation of the plasma column surface on the SFWs is more pronounced in comparison with the effect of the metal wall corrugation on SFWs propagating in such type of waveguides, which are partially filled with a plasma column, whose cross-section is of circular form.

## 5.4 Resonant Influence of Plasma-Dielectric Interface Shape on Surface Flute Wave Propagation in Magneto-Active Waveguides

As it was shown in the previous section, the frequency correction has a value of second order of smallness $\Delta\omega_n \propto \delta_n^2$, for SFWs propagating in waveguide structures with $N \neq 2\,|m|$. But the results obtained there are invalid in the resonant case, $N = 2\,|m|$. Degeneration of wave (or energetic) spectra happens in many branches of physics and its analysis has some specific features [28]. It was found in [29] that LF SFWs propagating in isotropic waveguides with non-circular plasma cross-section exist as two standings waves, whose frequencies are close. Because of these beating modes arise as the result of these standing waves interaction and the correction to the eigen frequency of SFWs becomes larger than it was in the non-resonant case considered in the previous section.

Here the waveguide structure with the same geometry and in the same framework as in Sect. 5.3 is studied. But unlike the case $N \neq 2\,|m|$ considered in the previous section let us study now just the case of SFW wave packets, whose fundamental harmonic satisfies the following resonant condition [30]: $|m| = N/2$. Then one can derive the secular dispersion relation $(D_0)^2 - (D_M^{(1)})^2 = 0$, which is typical for the case of isotropic waveguides. Moreover, if the value of the applied magnetic field is sufficiently small one can apply the theory developed for isotropic plasma filled waveguides in [29]. This statement is valid if the following inequality is satisfied:

$$\frac{|\omega_e|}{\Omega_e} \ll \frac{\delta_{2|m|}}{2}\sqrt{\frac{|m|}{\Delta}}\left[1 + \sqrt{\frac{\Omega_e a}{|m| c}}\,\varepsilon_d^{-1/4}\left(1 + \frac{a\Delta}{c}\Omega_e\right)\right]. \tag{5.25}$$

If the inequality (5.25) is not valid then one can analyze the influence of the waveguide cross-section shape on the dispersion properties of LF SFWs in the resonant case numerically. It was found that non-circularity of the plasma cross-section leads to a decrease of the SFW angular frequency if $m < 0$ (except for the case of the mode $m = -1$ propagating in a wide waveguide: $k_{ef} < 1$). The absolute value of $\Delta\omega_n$ is larger for the case of narrow waveguides than for wide waveguides (except for the case of the mode number $m = -1$, when it is very small), its value increases for decreasing magnitude of $B_0$, thickness and permittivity of dielectric layer. Changing the sign of the azimuthal mode number (from negative to positive) leads to changing sign of the frequency correction and also to some increase of its absolute value.

For the case $m = -1$ the results of numerical analyses of the LF SFW spectral composition are presented in Fig. 5.7. Curves calculated for the fast and slow satellites are marked by solid and dashed lines, respectively. Their analysis proves that the relative amplitudes of the slow and fast satellite harmonics sufficiently differ from each other unlike in the non-resonant case studied in the previous section. This can be explained by resonant increasing of the satellite harmonic, which propagates with the same angular phase velocity as the fundamental harmonic, but in the opposite direction (just this satellite harmonic is entitled here as resonant satellite harmonic). The magnitude of the relative amplitude of the resonant satellite harmonic essentially increases with increasing absolute value of the azimuthal mode number for the case $m < 0$. Decreasing the dielectric permittivity of the layer, which separates the plasma column from the waveguide metal wall, and the magnitude of the external magnetic field leads to an increase of the absolute value of the relative amplitude of the resonant satellite harmonic.

Numerical analysis of the spectral composition of the SFW wave packet, whose fundamental harmonic is characterized by $m < 0$ allows one to make the following conclusions on their peculiarities. First, the amplitudes of the fast satellites are lower than the amplitudes of the slow satellites. Second, the amplitudes of satellite harmonics are larger in wide waveguides as compared with the case of narrow waveguides. Third, the amplitudes of satellite harmonics decrease with decreasing value of $|m|$. Fourth, in the case of positive sign of the azimuthal mode number $m$, the

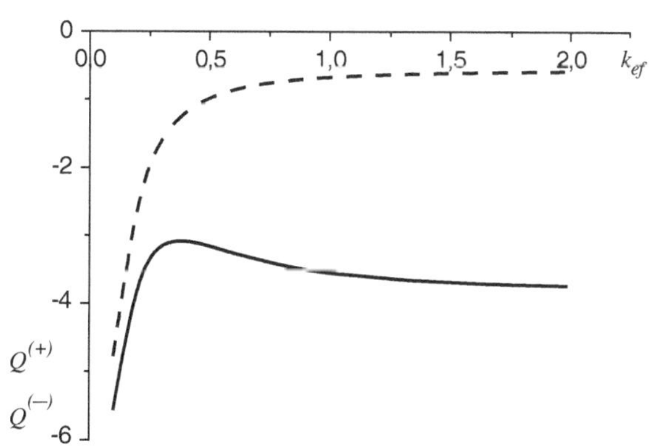


**Fig. 5.7** Amplitudes of satellite harmonics (the fast harmonic is depicted by solid line, the slow one is depicted by *dashed line*) versus $k_{ef}$ in the resonant case, $m = -1$, $z = 9$, $\varepsilon_d = 1$, $\Delta = 0.3$, $N = 2$

amplitude of the resonant satellite harmonic becomes firstly equal to zero and then becomes positive while the radius of the plasma column decreases until such waveguide becomes sufficiently small and can be already considered as a narrow one.

The possibility to control the SFW frequency spectrum and spectral composition of their wave packet by choosing the shape of the plasma column cross-section is shown in [30] for the resonant case. In this case the LF SFW frequency correction $\Delta\omega_n \propto B_0^{-1}$ and its value becomes much larger, $\Delta\omega_n \propto \delta_n$, than it was in the non-resonant case. In general, the influence of the plasma column cross-section non-circularity is stronger than in the case when LF SFWs propagate in waveguides, with longitudinally corrugated metal wall (see previous Sect. 5.2). This is explained by the fact that the amplitude of SFWs is maximal just on the plasma-dielectric interface and decreases if one goes away from the surface.

## 5.5 Influence of Plasma Column Cross-Section Non-Circularity on Dispersion Properties of High Frequency Surface Flute Waves Propagating in Magneto-Active Waveguides

Let us study the propagation of surface flute waves with extraordinary polarization in a perfectly conducting cylindrical metal chamber with radius $b$. In the chamber, there is a column of a cold plasma separated from the chamber wall by a dielectric layer with the permittivity $\varepsilon_d \geq 1$. An external uniform constant magnetic field is directed along the waveguide axis, $\vec{B}_0 || \vec{Z}$. It is assumed as well that the SFW propagates strictly transversely to the magnetic field and the waveguide is uniform along the cylinder axis, $\partial/\partial z \equiv 0$. The cross-section of the plasma column is described by expression (5.2). To illustrate the geometry of the plasma filled waveguide, which is under consideration in this section one can apply the Fig. 5.1 again, where line $R_2(\varphi)$, which describes the inner radius of the metal chamber must be changed to be a circle. The displacement vector and electric field vector are assumed to be related by the dielectric tensor of a weakly collisional cold magneto-active plasma (see, e.g., [1, 17, 18]). In contrast to the cases studied in the two previous sections, here the case of HF SFWs in the frequency range $\omega_{UH} < \omega < \omega_1$ is under consideration.

The dispersion properties of HF SFWs are investigated here by the method of successive approximations. In zero approximation, the plasma-dielectric interface is circular and is coaxial with the metal chamber. The properties of SFWs in this approximation can be described by using the results obtained in Chap. 2. When the radius of curvature of the plasma-dielectric interface is constant in the azimuthal direction, SFWs with different azimuthal mode numbers propagate independently of each other. That is why, in zero approximation, one can assume that only HF SFWs with the azimuthal mode number $m$ can propagate.

Because of the periodic spatial non-uniformity of the plasma-dielectric interface (see expression (5.2)), SFWs propagate as wave packets. Each such wave packet contains the fundamental harmonic, with a field varying as $\exp(im\varphi - i\omega t)$, and an infinite number of satellite harmonics, with fields varying as $\exp[im \pm jN)\varphi - i\omega t]$ ($j = 1, 2, 3\ldots$). From the theory of wave propagation in media whose properties vary periodically in one definite direction [24], the amplitudes of satellite harmonics are known to be small quantities (their values are of the order $(h_N)^j$) in comparison with the fundamental harmonic amplitude. With the aim of determining how the dispersion properties of SFWs depend on the shape of the noncircular cross-section of the plasma-dielectric interface, one can consider further a wave packet with only two nearest satellite harmonics, which are proportional to $\exp[i(m \pm N)\varphi - i\omega t]$. The reason for this approximation is that the higher satellite harmonics make only third and higher order contributions to the correction of the eigen frequency of SFWs. Using the expression (5.2) for the plasma-dielectric interface, one can see that, within second order terms, the boundary conditions (5.4) for the tangential components of the SFW fields take the form of continuity conditions at radius $r = a$, which are similar to expressions (5.20) and (5.21), respectively.

An analysis of the expressions for the SFW fields shows that the amplitudes of their satellite harmonics do not possess any symmetry: $H_z^{(+)} \neq H_z^{(-)}$. The spatial distribution of the azimuthal electric field and axial magnetic field of a SFW in a dielectric layer ($R_1 < r < b$) are expressed in terms of the Bessel functions of the first kind and the Neumann functions [25]:

$$E_\varphi^{(V)}(r,\varphi) = \frac{-i}{\sqrt{\varepsilon_d}}[S^{(0)}P_m'(\kappa r) + \sum_{\pm} S^{(\pm)}P_{m\pm N}'(\kappa r)\exp(\pm iN\varphi)]\exp(im\varphi - i\omega t), \tag{5.26}$$

$$H_z^{(V)}(r,\varphi) = [S^{(0)}P_m(\kappa r) + \sum_{\pm} S^{(\pm)}P_{m\pm N}(\kappa r)\exp(\pm iN\varphi)]\exp(im\varphi - i\omega t), \tag{5.27}$$

here $P_m(\kappa r) = J_m(\kappa r)N_m'(\kappa b) - J_m'(\kappa r)N_m(\kappa r)$; $\kappa^2 = \varepsilon_d k^2$; $S^{(0)}$ is a normalizing factor, which characterizes the amplitudes of the fundamental harmonics of the SFW fields in the dielectric layer. The prime index denotes the derivative of the function over its argument, symbol $\sum_{+}$ means summation over the upper and lower signs, the constants $S^{(+)}$ are determined by the boundary conditions (5.3)–(5.5), and the functions $P_m(\kappa r)$ and $P_{m\pm N}(\kappa r)$ are constructed in such a way that the azimuthal component of the electric field of the SFWs satisfies the boundary condition (5.5).

The radial dependence of the SFWs fields in the plasma region ($0 < r < R_1$) is expressed in terms of the modified Bessel functions and their derivatives [25]:

$$E_\varphi^{(p)}(r,\varphi) = \frac{ik}{k_\perp^2}\exp(im\varphi - i\omega t)\left\{A^{(0)}\left[\frac{m\mu}{r}I_m(k_\perp r) + k_\perp I_m'(k_\perp r)\right] + \right.$$
$$\left. + \sum_{\pm} A^{(\pm)}\left[\frac{(m \pm N)\mu}{r}I_{m\pm N}(k_\perp r) + k_\perp I_{m\pm N}'(k_\perp r)\right]\exp(\pm iN\varphi)\right\} \tag{5.28}$$

$$H_z^{(p)}(r,\varphi) = \exp(im\varphi - i\omega t)[A^{(0)} I_m(k_\perp r) + \sum_{\pm} A^{(\pm)} I_{m\pm N}(k_\perp r) \exp(\pm iN\varphi)], \tag{5.29}$$

here $A^{(0)}$ is a normalizing factor, which characterizes the amplitude of the fundamental harmonics of the SFW fields in the plasma column. The constants $A^{(\pm)}$, which characterize the amplitudes of the satellite harmonics of these fields in the plasma, can be determined from the boundary conditions indicated above. The expressions (5.28) and (5.29) for the field components $E_\varphi^{(p)}(r,\varphi)$ and $H_z^{(p)}(r,\varphi)$ satisfy the boundary condition (5.3).

In first order approximation, using the indicated boundary conditions one can derive relations between the amplitudes of the satellite harmonics of the HF SFW fields and the amplitudes of their fundamental harmonics for the vacuum and plasma regions, respectively. They are given by the following expressions:

$$A^{(\pm)} = p_1^{(\pm)} A^{(0)} + q_1^{(\pm)} S^{(0)}, \quad S^{(\pm)} = p_2^{(\pm)} A^{(0)} + q_2^{(\pm)} S^{(0)}, \tag{5.30}$$

$$p_1^{(\pm)} = (\alpha_2^{(\pm)} \beta_3^{(\pm)} - \alpha_3 \beta_2^{(\pm)})/D^{(\pm)}, \quad p_2^{(\pm)} = (\alpha_3 \beta_1^{(\pm)} - \alpha_1^{(\pm)} \beta_3^{(\pm)})/D^{(\pm)}, \tag{5.31}$$

$$q_1^{(\pm)} = (\alpha_2^{(\pm)} \beta_4^{(\pm)} - \alpha_4 \beta_2^{(\pm)})/D^{(\pm)}, \quad q_2^{(\pm)} = (\alpha_4 \beta_1^{(\pm)} - \alpha_1^{(\pm)} \beta_4^{(\pm)})/D^{(\pm)}, \tag{5.32}$$

$$D^{(\pm)} = \alpha_1^{(\pm)} \beta_2^{(\pm)} - \alpha_2^{(\pm)} \beta_1^{(\pm)}, \tag{5.33}$$

$$\alpha_1^{(\pm)} = I_{m\pm N}(k_\perp a), \quad \alpha_2^{(\pm)} = -P_{m\pm N}(\kappa a), \tag{5.34}$$

$$\alpha_3 = 0.5 h_N k_\perp a I_m'(k_\perp a), \quad \alpha_4 = -0.5\, h_N \kappa\, a P_m'(\kappa\, a), \tag{5.35}$$

$$\begin{aligned} \beta_1^{(\pm)} &= k k_\perp^{-2} \left[ (m \pm N)\, \mu\, I_{m\pm N}(k_\perp a)/a + k_\perp I_{m\pm N}'(k_\perp a) \right], \\ \beta_2^{(\pm)} &= P_{m\pm N}'(\kappa\, a)/\sqrt{\varepsilon_d}, \end{aligned} \tag{5.36}$$

$$\beta_3^{(\pm)} = -0.5\, h_N k k_\perp^{-2} \left[ \left( m\mu \mp mN - k_\perp^2 a^2 - m^2 \right) I_m(k_\perp a)/a + k_\perp I_m'(k_\perp a)(1 - m\mu \mp \mu N) \right], \tag{5.37}$$

$$\beta_4^{(\pm)} = 0.5 h_N k a^{-1} \kappa^{-2} \left[ \left( m^2 \pm mN - a^2 \kappa^2 \right) P_m(\kappa\, a) + \kappa\, a P_m'(\kappa\, a) \right]. \tag{5.38}$$

The dispersion relation for HF SFWs propagating along the corrugated plasma interface has the same form as it was obtained in [26] for LF SFWs (see Sect. 5.3)

$D_0 + D_N^{(2)} = 0$, where $D_0(\omega_0) = 0$ is dispersion equation of SFXMs in zero approximation (see Sect. 2.1). Thus let us write down here only the expression for $D_N^{(2)}$ [super-script (2) means that it is a second order correction over the small parameter of the problem]:

$$D_N^{(2)} = \alpha_1^{(0)}\beta_2^{(2)} - \alpha_2^{(2)}\beta_1^{(0)} + \alpha_1^{(2)}\beta_2^{(0)} - \alpha_2^{(0)}\beta_1^{(2)}, \tag{5.39}$$

here

$$\begin{aligned}
\alpha_1^{(2)} &= h_N a\{k_\perp[p_1^{(+)} I'_{m+N}(k_\perp a) + p_1^{(-)} I'_{m-N}(k_\perp a)] \\
&\quad - \kappa[p_2^{(+)} P'_{m+N}(\kappa a) + p_2^{(-)} P'_{m-N}(\kappa a)]\}/2 \\
&\quad + 0.25\, h_N^2 k_\perp^2 a^2 I''_m(k_\perp a); \\
\alpha_2^{(2)} &= h_N k_\perp a\,[q_1^{(+)} I'_{m+N}(k_\perp a) + q_1^{(-)} I'_{m-N}(k_\perp a)]/2 \\
&\quad -0.5\, h_N \kappa\, a\,[q_2^{(+)} P'_{m+N}(\kappa a) + q_2^{(-)} P'_{m-N}(\kappa a)] - 0.25 h_N^2 \kappa^2 a^2 P''_m(\kappa a);
\end{aligned}$$

$$\begin{aligned}
\beta_1^{(2)} &= 0.5 h_N N\,[p_1^{(-)}\beta_5^{(-)} - p_1^{(+)}\beta_5^{(+)}] + 0.5 a h_N [p_1^{(-)}\beta_6^{(-)} + p_1^{(+)}\beta_6^{(+)}] + 0.25 h_N^2 a^2 \beta_7 \\
&\quad + 0.5 h_N N k \kappa^{-2} a\,[(m-N) p_2^{(-)} P_{m-N}(\kappa a) - (m+N) p_2^{(+)} P_{m+N}(\kappa a)] \\
&\quad + 0.5 a k h_N [p_2^{(-)} P''_{m-N}(\kappa a) + p_2^{(+)} P''_{m-N}(\kappa a)];
\end{aligned}$$

$$\begin{aligned}
\beta_2^{(2)} &= h_N[N(q_1^{(-)}\beta_5^{(-)} - q_1^{(+)}\beta_5^{(+)}) + a(q_1^{(-)}\beta_6^{(-)} + q_1^{(+)}\beta_6^{(+)})]/2 \\
&\quad + h_N^2 a^2 \kappa^2 P'''_m(\kappa a)/(4\sqrt{\varepsilon_d}) \\
&\quad + h_N N k\,[(m-N) q_2^{(-)} P_{m-N}(\kappa a) - (m+N) q_2^{(+)} P_{m+N}(\kappa a)]/(2 a \kappa^2) \\
&\quad + 0.5 a k h_N [q_2^{(-)} P''_{m-N}(\kappa a) + q_2^{(+)} P''_{m-N}(\kappa a)];
\end{aligned}$$

$$\beta_5^{(\pm)} = \frac{k}{k_\perp^2}\left[\frac{(m \pm N)}{a} I_{m\pm N}(k_\perp a) + \mu k_\perp I'_{m\pm N}(k_\perp a)\right];$$

$$\beta_6^{(+)} = \frac{k}{k_\perp^2}\left[\frac{(m \pm N)\mu}{-a^2} I_{m\pm N}(k_\perp a) + \frac{(m \pm N)\mu}{a} k_\perp I'_{m\pm N}(k_\perp a) + k_\perp^2 I''_{m\pm N}(k_\perp a)\right];$$

$$\beta_7 = \frac{k}{k_\perp^2}\left[\frac{2m\mu}{a^3} I_m(k_\perp a) - \frac{2m\mu}{a^2} k_\perp I'_m(k_\perp a) + \frac{m\mu}{a} k_\perp^2 I''_m(k_\perp a) + k_\perp^3 I'''_m(k_\perp a)\right].$$

The solution of the dispersion relation for HF SFWs can be represented in the form $\omega = \omega_0 + \Delta\omega_N$, where $\omega_0$ is the eigen angular frequency of the HF SFXMs (see Sect. 2.1). The frequency correction $\Delta\omega_N$ can be calculated for any index $N$ according to formula (5.12) like it was done in Sect. 5.1.

With an accuracy up to the second order of smallness over the small parameter $\delta_n$ the different small terms $\delta_n \cos(n\varphi + \varphi_{1n})$ in expression (5.2) for plasma column radius make their contribution independently to the correction of the HF SFW frequency. That is why the dependence of the eigen frequency of these waves on the deviation of the cross-section of the plasma-dielectric interface from circular shape can be described by representing the complete correction $\Delta\omega$ to the eigen frequency as the sum of particular corrections $\Delta\omega_n$, which can be calculated according to formula (5.12).

Application of a thin dielectric gap, which separates the plasma column from the metal chamber wall, causes a shift of the dispersion curve $\omega = \omega\,(k_{ef})$ of HF SFWs towards greater values of the effective wave number $k_{ef} = |m|\,\delta/a$ within the considered frequency range, unlike in the case considered in Sect. 5.1:

$$\omega_0 \approx \sqrt{\Omega_e^2(1 + k_{ef}^2) + \omega_e^2} - |m|\,\omega_e\Delta(\varepsilon_d - k_{ef}^2). \tag{5.40}$$

Nearby the lower limits of the HF range, the SFW penetration depth into the plasma becomes small, so that inequality $k_\perp a \gg 1$ becomes valid. This means that in this case $\varepsilon_1 \to 0$. As far as a dispersion curve of the HF SFW asymptotically approaches to the lower range limit for smaller values of effective wave number ($k_{ef} \ll 1$) then the derivative, which is located in the denominator of the expression for the SFW frequency correction $\Delta\omega_n$ goes to infinity, or in other words $(\partial D_0/\partial\omega)^{-1}_{|\omega=\omega_0} \to 0$. Therefore, in this case, one can expect a vanishing effect of the deviation of the cross-section of the plasma-dielectric interface from circular shape on dispersion properties of HF SFWs. It should be mentioned that in this band of $k_{ef}$ one does not observe effective excitation of HF SFWs by gyrating electron beams [31]. This property of HF SFWs makes them distinctive from LF SFWs, whose frequency is almost proportional to $k_{ef}$ in wide waveguides.

Unlike that, nearby the upper HF range limit (or, in other words, nearby the right limit of the $k_{ef}$ band, within which HF SFWs can propagate), SFWs well penetrate into the plasma, $k_\perp a \gg 1$, which means validity of the relation $\varepsilon_1 \approx \varepsilon_2$. Nearby the upper HF range limit, the frequency correction of SFWs, caused by the deviation of the cross-section of the plasma-dielectric interface from circular shape, strongly increases $\Delta\omega_N \propto k_\perp^{-2}$. Thus, the usability condition, $\Delta\omega_N \ll \omega_0$, of the method of successive approximations applied here is violated.

The results of numerical investigations of the waveguide cross-section non-circularity on dispersion properties and spatial distribution of HF SFWs are illustrated in Figs. 5.8, 5.9, 5.10, 5.11, 5.12 and 5.13. The figures are given in terms of dimensionless parameters, the correction to the angular frequency of HF SFWs, $\Delta\omega_N/(\Omega_e h_N^2)$, normalized to the electron plasma frequency and the squared small parameter of the problem versus the effective wave number $k_{ef}$. The electron plasma frequency can conveniently be chosen as a normalizing divisor for the following two reasons. First, the plasma density and, accordingly, the plasma frequency are constant in the model under consideration. Second, in the plots, one can immediately see how the correction to the eigen frequency of SFWs scales with such a characteristic parameter like the electron plasma frequency. The choice of the skin depth $\delta$ (penetration depth of electromagnetic fields into the plasma) as a normalizing factor

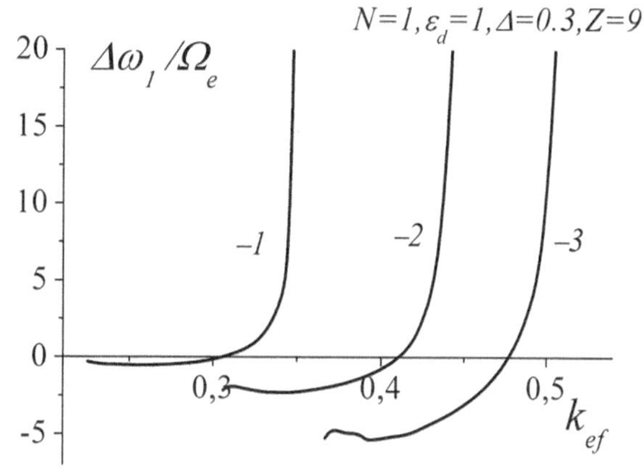


**Fig. 5.8** HF SFWs frequency correction versus $k_{ef}$; N = 1; Z = 9; $\varepsilon_d = 1$; $\Delta = 0.3$; $m = -1, -2, -3$

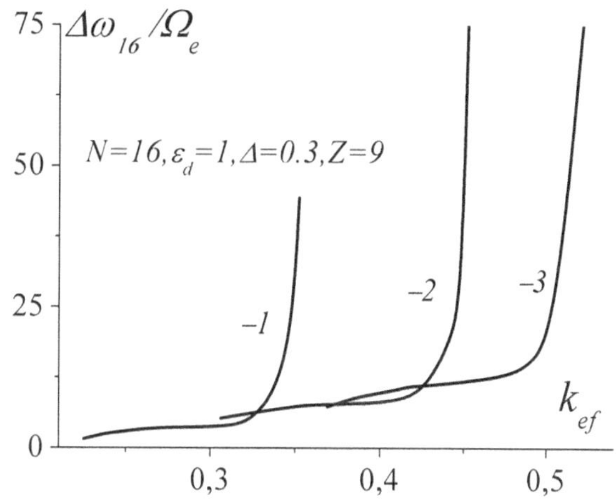


**Fig. 5.9** The same as in Fig. 5.8, but for N = 16

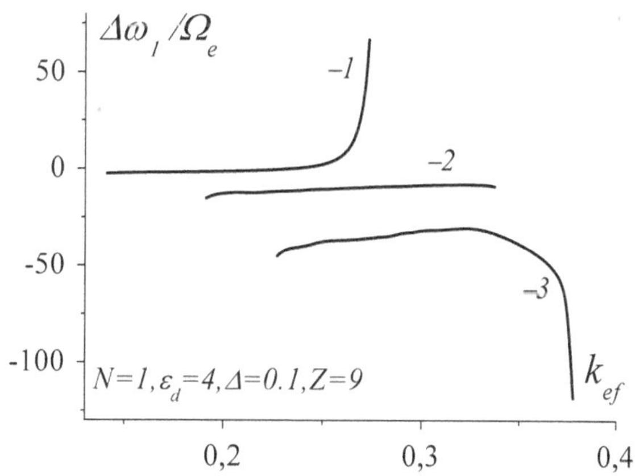


**Fig. 5.10** The same as in Fig. 5.8, but for $\varepsilon_d = 4$; $\Delta = 0.1$

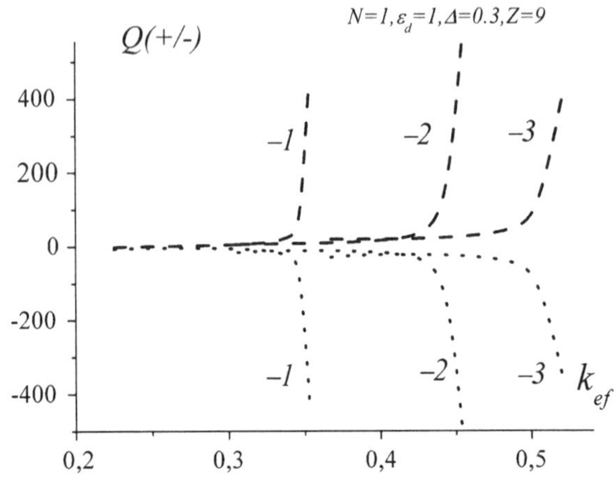


**Fig. 5.11** Amplitudes of satellite harmonics versus $k_{ef}$. The numbers indicate the values of the azimuthal mode numbers. Cases of slow and fast harmonics are depicted by dashed and dotted lines, respectively. Plasma waveguide parameters are the same as in Fig. 5.8

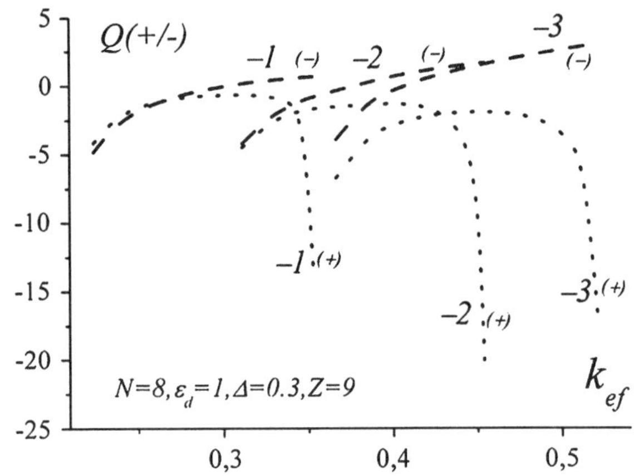


**Fig. 5.12** The same as in the Fig. 5.11, but for N = 8

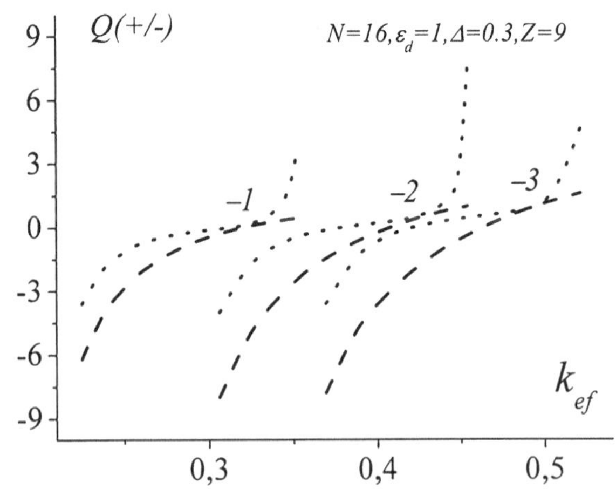


**Fig. 5.13** The same as in the Fig. 5.11, but for N = 16

appears to be convenient when using the fluid approximation. The numbers in these figures denote the azimuthal mode number $m$.

Let us compare the case under consideration with the case of LF SFWs. Note, that the deviation of the cross-section of the plasma-dielectric interface from circular shape causes a decrease of the eigen frequency of LF SFWs with negative mode numbers [except for the mode $m = -1$, if they propagate in wide waveguides ($k_{ef} \ll 1$)]. Thus a cylindrical metal waveguide with non-circular plasma-dielectric interface is a slowing-down structure for such waves. On the other hand, the frequency correction for HF SFWs caused by the deviation of the cross-section of the plasma-dielectric interface from circular shape, in its turn, can be both negative and positive depending on the values of the waveguide parameters.

Figure 5.8 shows how the frequency correction $\Delta\omega_N/(\Omega_e h_N^2)$, depends on $k_{ef}$ for $N = 1 \leq |m|$, $\varepsilon_d = 1$, $\Delta = 0.3$, $Z = \Omega_e/|\omega_e| = 9$. Analyzing the curves plotted in this figure one can see that the minimum of the frequency correction $\Delta\omega_1$ strongly depends on the mode number $\Delta\omega_1 \propto |m|^\beta$, where $\beta \approx 2.102$. The dependence of $\Delta\omega_N/(\Omega_e h_N^2)$ on $k_{ef}$ is shown for the case $N = 16 > |m|$ in Fig. 5.9. Increase in $N$ leads to reduction of the wavebands of $k_{ef}$, where the frequency corrections $\Delta\omega_N$ are negative. For $N = 8$, the frequency correction for SFWs with $m = -1$ is positive for all the range of $k_{ef}$ under the consideration, whereas the curves $\Delta\omega_N(k_{ef})$ for the SFWs with $m = -2, -3$ still contain the bands of $k_{ef}$, where the corrections are negative. For $N = 12$ the frequency corrections are positive already for all three values of the azimuthal mode number. Increasing $N$ is not followed by any essential increase in the absolute value of $\Delta\omega_N$. An analysis of the expression for $\Delta\omega_N(k_{ef})$ shows that it converges if the small parameters $h_n$ decrease with the index of summation $n$ faster than $n^{-1.5}$. This requirement is not too restrictive for our study because, when, e.g., the plasma-dielectric interface has a square cross-section, the small parameters of the corrugation $h_n$ decrease inversely to the square of the summation index $n$:

$$h_n = 4\sqrt{2}(-1)^{\frac{n}{4}}/\left[n^2 \ln\left(3+2\sqrt{2}\right)\right]. \tag{5.41}$$

Increasing $B_0$ leads to an essential decrease of the correction $\Delta\omega_1$. It should be underlined that the dispersion properties of SFWs, which propagate in magnetoactive waveguides, differ strongly from the properties of SFWs propagating in non-magnetized waveguides. For instance, if an external magnetic field is applied SFWs can propagate along a cold plasma-metal interface [27]. This is impossible in the case of $B_0 = 0$. Let us also point out that the spectra of SFWs are degenerate with respect to the sign of the azimuthal mode number in circular plasma waveguides without constant magnetic field: $\omega^{(0)}(+|m|) = \omega^{(0)}(-|m|)$.

However, in a waveguide filled with cold plasma, whose interface has a non-circular cross-section, SFWs with $n = 2|m|$ are generally non-degenerate [29]. Let us remind for comparison, that even in a weak ($\Omega_e^2 \gg \omega_e^2\varepsilon_d$) magnetic field, the dispersion properties of LF SFWs with these azimuthal mode numbers ($|m| \neq n/2$) are more sensitive to the shape of the cross-section of the plasma-dielectric interface than those of LF SFWs with other azimuthal mode numbers ($|m| \neq n/2$), for the same other conditions [30]. The reason is, that for LF SFWs with azimuthal mode numbers $|m| \neq n/2$, the denominators $D^{(\pm)}$ in expressions (5.31) and (5.32) for $p_1^{(\pm)}$, $p_1^{(\pm)}$, $q_1^{(\pm)}$, and $q_1^{(\pm)}$ are very small. Namely, if $m > 0$, the denominator $D^{(+)}$ is not small, but the denominator $D^{(-)}$ is:

$$D^{(-)} = -P_m(\kappa\, a)\frac{2m\mu k}{k_\perp^2 a} I_m(k_\perp a) \ll D^{(+)}. \tag{5.42}$$

And vice versa: if $m < 0$ the denominator $D^{(+)}$ is small [in this case it is expressed by the Eq. (5.33)], but $D^{(-)}$ is not of small value. It should be reminded that in the HF range, SFWs can propagate in waveguides with narrow dielectric layer only with negative azimuthal mode numbers. That is why the deviation of the cross-section of the plasma-dielectric interface from circular shape does not cause any resonant influence on the dispersion properties of HF SFWs. Moreover, if there is no external magnetic field the HF range does not exist at all.

The influence of the dielectric layer thickness $\Delta$ on the behavior of the curves representing the dependence of the frequency correction on $k_{ef}$ appeared to be not essential, unlike in the case of LF SFWs [26], for which $\Delta\omega_1 \propto \Delta^{-2}$ with high accuracy. It also should be mentioned, that an infinite increase of frequency correction $\Delta\omega_1$ value starts from smaller values of the parameter $k_{ef}$, if $\Delta$ decreases.

Figure 5.10 illustrates the influence of the dielectric permittivity of the dielectric layer, which separates the plasma column from the metal wall of the chamber, on the frequency correction $\Delta\omega_1/(\Omega_e h_1^2)$. Here the following parameters of the waveguide were chosen: $N = 1$, $\Delta = 0.1$, $Z = 9$, dielectric permittivity $\varepsilon_d = 4$. An increase of the $\varepsilon_d$ value not only strongly effects the absolute value of $\Delta\omega_1$, but also changes its sign, which determines the limit of the frequency corrections nearby the right band limit of $k_{ef}$, where HF SFWs with given azimuthal mode number exist. For instance, for HF SFWs with azimuthal mode number $m = -3$, the normalized frequency correction is approximately equal to $\Delta\omega_1/(\Omega_e h_1^2) \approx -200$. To satisfy the usability condition of the method of successive approximations, $\Delta\omega_1/\omega_0 \ll 1$, the small parameter of the problem should be sufficiently small, namely, the following rigid condition should be valid: $h_1 \ll 0.07$.

Figure 5.11 shows how the relative amplitudes $Q^{(\pm)}$ of the satellite harmonics, varying as $\exp[i(m \pm N)\varphi - i\omega t]$, depend on the effective wave number,

$$Q^{(\pm)} = A^{(\pm)} I_{m\pm N}(k_\perp a)/[h_1 A^{(0)} I_m(k_\perp a)]. \tag{5.43}$$

The calculations were carried out for the same plasma waveguide parameters as in Fig. 5.7. For the wave packet of SFWs with azimuthal mode number $m = -3$ the fundamental harmonic and its two satellite harmonics propagate in the same direction. The angular phase velocities of these satellite harmonics $\omega/(m \pm N)$ are equal to $(-\omega_0/2)$ and $(-\omega_0/4)$, respectively; i.e., one of the satellite harmonics, varying as $\propto \exp[i(m + N)\varphi - i\omega t]$, propagates faster than the fundamental harmonic, whose angular phase velocity is equal to $-\omega_0/3$, and the other satellite harmonic, varying as $\propto \exp[i(m - N)\varphi - i\omega t]$, propagates slower than the main one. For the wave packet of SFWs with the fundamental harmonic $m = -1$, one of the satellites, varying as $\propto \exp[i(m+N)\varphi - i\omega\, t]$, is a standing wave in azimuthal direction. The other satellite, varying as $\propto \exp[i(m-N)\varphi - i\omega\, t]$, has the angular phase velocity $-\omega_0/2$; i.e., it propagates slower than the fundamental harmonic, which has the angular phase velocity $-\omega_0$.

Unlike the LF SFWs, whose amplitudes of the fast satellite harmonics are lower than the amplitudes of the slow satellite harmonics [26], the amplitudes of HF SFW satellite harmonics are approximately symmetric: $Q^{(+)} \approx Q^{(-)}$ in the case of waveguides with $N = 1$. This symmetry does not change with changing the parameters $\Delta$, $Z$ and $\varepsilon_d$. For wide waveguides (if $k_{ef} \to 0$) the amplitudes of satellite harmonics of HF SFWs turn to zero which confirms the usability condition of the method of successive approximations. But approaching the right limit of the $k_{ef}$ range, where HF SFWs with a given value of $m$ can exist, the amplitudes of HF SFWs satellite harmonics become unrestrictedly large and the usability condition of the method of successive approximations, $Q^{(\pm)} \sim h_N \ll 1$, is violated. In the case of $N = 1$, the absolute values of $Q^{(\pm)}$ increase with increasing $Z$ and $\varepsilon_d$, but weakly depend on the thickness $\Delta$. We do not present the plots of the dependencies of SFW satellite harmonics amplitudes calculated for the values of the waveguides parameters, which correspond to Figs. 5.9 and 5.10, since they do not differ qualitatively from those presented in Fig. 5.11.

For satellite harmonics in the case $N \gg |m|$, the character of the dependence of $Q^{(\pm)}$ on the effective wave number differs from that in the case of $N = 1 \le |m|$. Plots of the dependence of $Q^{(\pm)}$ on $k_{ef}$ calculated for the same values of the waveguides parameters as done in Fig. 5.8, but for the satellite harmonics with $N = 8$, 16 are presented in Figs. 5.12 and 5.13, respectively. The curves calculated for the amplitudes $Q^{(-)}$ of slow satellite harmonics, which propagate in the same direction as the fundamental harmonic, are shown by dashed lines. The curves which illustrate the dependencies of the satellite harmonics amplitudes $Q^{(+)}$, which propagate in the opposite direction as compared to the fundamental harmonics, are shown by dotted lines. Analyzing these dependences, let us note, first of all, the asymmetry of the dependencies of $Q^{(\pm)}$. The amplitudes $Q^{(-)}$ are lower than $Q^{(+)}$. In the case $N = 8$, the amplitude $Q^{(-)}$ reaches its maximum absolute value 4.82 (see Fig. 5.12) already nearby the left limit of the $k_{ef}$ range, where SFW modes with $m = -1$ can exist. Increasing number $N$ leads to widening the $k_{ef}$ ranges, where the derivative $\partial Q^{(+)}/\partial k_{ef}$ is positive. In the case $N = 16$ (see Fig. 5.13), the amplitudes $Q^{(+)}$ of all first three surface flute modes turn to positive infinity with approaching of $k_{ef}$ to the right limits of the $k_{ef}$ ranges, where SFWs can exist.

Let us employ the above obtained results for estimating how the dispersion properties of SFWs depend on the non-coincidence between the axes of the plasma cylinder and the metal waveguide (that means $N = 1$). For example, for a laboratory gaseous plasma with a density of $n = 10^{11}$ cm$^{-3}$, the electron plasma frequency is equal to $1.8 \times 10^{10}$ s$^{-1}$ and the skin depth is about 1.7 cm. The relationship $\Omega_e = 9|\omega_e|$ between electron plasma frequency and electron cyclotron frequency can be achieved with a moderate magnetic field, $B_0 \approx 0.0161\ T$. For a SFW with azimuthal mode number $m = -1$, which propagates in a plasma cylinder of radius a $\approx$ 8.5 cm, the effective wave number is equal to $k_{ef} = 0.2$.

The condition $\Delta = 0.1$ in this case implies that the thickness of the vacuum layer is about $b - a \approx 0.85\,\text{cm}$. This is less than 9 % of the vacuum wavelength, $(b-a)\omega/c \approx 0.52$, so the usability condition for the method of successive approximations is satisfied. In this case a SFW propagates at the frequency 2.9 GHz, which is 1.02 times higher than the electron plasma frequency. If the axes of the plasma column and the metal waveguide are separated by a distance of 1 mm, the small parameter of the problem is approximately equal to $h_1 \approx 0.012$. In this case, the correction to the SFW frequency is as small as 662 kHz, which is much less than the eigen frequency, as it is required by the method of successive approximations. In such a waveguide, the amplitude of the satellite harmonic with the azimuthal mode number $m = -2$ is lower than 2.4 % of the amplitude of the fundamental harmonic. The amplitude of the standing harmonic in the azimuthal direction (which means validity equation $m = 0$) is somewhat larger in such a waveguide: it is equal to 2.9 % of the fundamental harmonic amplitude.

## 5.6 Conclusions

This Chapter is devoted to the theory of SFWs propagating in magneto-active plasma filled waveguides with non-circular cross-section. Periodic change of the curvature of the plasma column interface with a dielectric layer and/or a metal chamber along the azimuthal angle is shown to cause the propagation of electromagnetic surface waves in the form of wave packets. As a rule, the SFW eigen frequency correction, which is caused by the deviation of the waveguide cross-section from circular shape, is found to be a small value of second order over the small parameter of the problem. In the special (resonant) cases associated with degeneration of the frequency spectrum, the value of the frequency correction for the LF SFWs can be proportional to the small parameter of the problem in the first power.

Periodic azimuthal non-uniformity of the plasma waveguide is shown to result in splitting the spectra of eigen SFWs, if the angular period of the non-uniformity of the dielectric layer thickness is equal to half the angular period of the wave. The value of this splitting is found to be of first order of smallness. The correction to the SFW eigen frequency, caused by the non-uniformity of the dielectric layer thickness, is more sensitive to a corrugation of the plasma-dielectric layer interface (or its deviation from circular shape) than to a corrugation of the metal chamber. Thus designing a waveguide with a special shape of the cross-section can be used for control over the eigen frequency and spectral contents of the electromagnetic eigen waves propagating there.

The results presented here can be essentially interesting both for studying wave processes in fusion devices and for plasma electronics. For example, corrugated waveguide lines are utilized as interaction circuits of free electron masers [32]. One can strongly enhance the power (about 20 times) of HF electromagnetic radiation [33] obtained by the aid of sectioned waveguide-compressors with screw corrugated surface. This is associated with the well-known advantages of plasma waveguides [2, 4, 34] in comparison with vacuum waveguides, namely: enhanced value of limiting current, optimized conditions of beam-wave interaction, possibility to enhance the power of the generated signal and to regulate smoothly its frequency. The applicability of the method of successive approximation has been proved not only analytically in theoretical papers referenced here, but there is also an experimental confirmation [35] of its sufficient accuracy and usability for practical calculations.

The frequency correction of HF SFWs propagating in waveguides, which are partially filled with plasma, is shown to increase with decreasing external magnetic field and increasing permittivity of the dielectric layer. Varying the thickness of the dielectric layer (under the initial assumption that this layer is thin) weakly affects the frequency correction. The sign and absolute value of $\Delta\omega_N$ are very sensitive to the value of the angular period of the corrugation of the plasma-dielectric interface (parameter $N$).

Examination of the spectral content of HF SFWs propagating along the surface of a plasma column with non-circular cross-section makes it possible to reveal the following properties of the wave packet. In the case $N = 1$, the amplitudes of the satellite harmonics of HF SFWs are almost anti-symmetric, $Q^{(-)} \approx -Q^{(+)}$. In the case $N \gg |m|$, the amplitudes $Q^{(-)}$ are small compared with those of $Q^{(+)}$.

## References


1. Alexandrov, A. F., Bogdankevich, L. S., & Rukhadze, A. A. (1984). *Principles of plasma electrodynamics*. Berlin and New York: Springer.
2. Kuzelev, V. M., & Rukhadze, A. A. (1990). *Electrodynamics of dense electron beams in plasma*. Moscow: Nauka. (in Russian).
3. Miller, R. (1982). *Introduction to the physics of strong-current beams of charged particles*. New York: Plenum Press.
4. Kondratenko, A. M., & Kuklin, V. M. (1988). *Basis of plasma electronics*. Moscow: Energoatomizdat. (in Russian).
5. Landau, L. D., & Lifshits, Ye.M. (1960). *Course of theoretical physics. Electrodynamics of continuous media* (Vol. 8). Oxford: Pergamon Press.
6. Gradov, O. M., & Stenflo, L. (1983). Linear theory of a cold bounded plasma. *Physics Reports (Review Section Physics Letter)*, *94*, 111–137.
7. Azarenkov, N. A., & Ostrikov, K. N. (1991). Surface magnetoplasma waves at the interface between a plasma-like medium and a metal in a Voigt geometry. *Physics Reports, 308*, 333–428.
8. Beletski, N. N., Bulgakov, A. A., Khankina, S. I., & Yakovenko, V. M. (1984). *Plasma instabilities and non-linear phenomena in semi-conductors*. Kyiv: Naukova dumka. (in Russian).
9. Gradov, O. M., & Stenflo, L. (2001). Theory of non-linear plasma surface waves. *Journal of Plasma Physics, 65*, 73–77.
10. Wesson, J. A., & Lashmore-Davies, C. N. (1997). *Tokamaks*. Oxford: Clarendon Press.
11. Volkov, E. D., Suprunenko, V. A., & Shyshkin, O. O. (1983). *Stellarators*. Kyiv: Naukova dumka. (in Russian).
12. Beidler, C. D., Grieger, G., Harmeyer, E., et al. (1995) *Helias reactor studies*. Preprint of Max-Planck-Institut für Plasmaphysik, 85748 Garching bei München, FRG, *IPP 2/330*.

13. Ballico, R. C., & Cross, M. J.(1990). Probe measurements of ICRF Alfven surface waves in the TORTUS tokamak. *Fusion Engineering and Design*, *12*(1–2), 197–201.
14. Murphy, A. B. (1990). Waves in the edge plasma during ion cyclotron resonance heating. *Fusion Engineering and Design*, *12*(1–2), 79–92.
15. Porkolab, M. (1990). Parametric instabilities in the tokamak edge plasma in the ion cyclotron heating regimes. *Fusion Engineering and Design*, *12*(1–2), 93–103.
16. Noterdaeme, J.-M., ICRH-team, ASDEX-team, & NI-team. (1990). Experimental results on edge effects during ICRF heating of ASDEX plasmas. *Fusion Engineering and Design*, *12*(1–2), 127–201.
17. Akhiezer, A. I., Akhiezer, I. A., Polovin, R. V., Sitenko, A. G., & Stepanov, K. N. (1975). *Plasma electrodynamics*. Oxford: Pergamon Press.
18. Krall, N. A., & Trivelpiece, A. W. (1973). *Principles of plasma physics*. NY: MacGraw-Hill.
19. Azarenkov, M. O., & Kondratenko, A. M. (1989). Surface waves in gyro-tropic plasma layers bounded with metal. *Soviet Journal of Communications Technology and Electronics*, *34*, 1525–1527 (in Russian).
20. Girka, I., & Zolotukhin, O. (1994). Transversal surface magneto-plasma waves in metal waveguide of rectangular cross-section filled with n-semiconductor. *Soviet Journal of Communications Technology and Electronics*, *39*(1–2), 1961–1968.
21. Agranvich, V. M., & Mills, D. L. (1982). *Surface polaritons*. Amsterdam: North-Holland.
22. Girka V., Girka I., Girka A., et al. (2011). Theory of azimuthal surface waves propagating in non-uniform waveguides. *Journal of Plasma Physics*, *77*(4), 493–519.
23. Girka, I., Girka, V., & Pavlenko, I. (1997). Surface waves propagating in the direction transverse to the axis of magnetized plasma—filled waveguides with noncircular transverse cross sections. *Plasma Physics Reports*, *23*(11), 959–963.
24. Karpov, S. Y., & Stolyarov, S. N. (1993). Propagation and transformation of electromagnetic waves in one-dimensional periodic structures. *Physics-Uspekhi*, *36*(1), 1–22.
25. Abramowitz, M., & Stegun I. A. (Eds.). (1964). *Handbook of mathematical functions*. National Bureau of Standards, Applied Mathematics, Series 55, NY.
26. Girka, I. O., Girka, O. I., Girka, V. O., et al. (2007). Effect of the shape of the cross- section of a plasma-dielectric interface on the dispersion properties of azimuthal surface modes. *Plasma Physics Reports*, *33*(2), 91–101.
27. Girka, V., Girka, I., Pavlenko, I., et al. (2012). Coupled azimuthal modes propagating in current-carrying plasma waveguides. *Journal of Plasma Physics*, *78*(2), 105–123.
28. Landau, L. D., & Lifshits, E. M. (1977). *Course of theoretical physics. quantum mechanics* (Vol. 3). Oxford: Pergamon Press.
29. Girka, I. O. (1997). Multi-peak structure of the spectra of azimuthal surface waves in isotropic plasma waveguides with noncircular transverse cross-sections. *Plasma Physics Reports*, *23*(3), 246–251.
30. Girka, I., Girka, O., Girka, V., et al. (2007). Resonant effect of the noncircular shape of the plasma surface on the dispersion properties of extraordinary azimuthal surface modes in magneto-active waveguides. *Plasma Physics Reports*, *33*(7), 543–552.
31. Girka, V., Girka, I., & Pavlenko, I. (2011). Excitation of azimuthal surface modes by relativistic flows of electrons in the high-frequency range. *Plasma Physics Reports*, *37*(5), 447–454.
32. Ginzburg, N. S., Zaslavskii, V. Y., Peskov, N. Y., et al. (2006) Theory of a planar free-electron maser with transverse electromagnetic flux circulation in a 2D Bragg mirror. *Technical Physics*, *51*(12), 1618–1623.
33. Mishakin, S. V., & Samsonov, S. V. (2009). Optimization of frequency-modulated pulse compression in a sectioned waveguide with a helically corrugated surface. *Technical Physics*, *79*(11), 1655–1662.
34. Onishchenko, I. M., Sydorenko, D. Y., & Sotnykov, G. V. (1995). Electron beam interaction with a plasma-filled rippled coaxial waveguide. *Plasma Physics Reports*, *21*(8), 670–676.
35. Aksenova, L. A., Kejer, A. P., Nemirovskaja, L. L., et al. (1988) Experimental investigation of mode transformation in waveguides with corrugated walls. *Izvestija Vuzov Radiofizika*, *31*, 1223 (in Russian).

# Chapter 6
# Applications of Surface Wave Propagation

This chapter is devoted to considerations on some possible applications of the surface waves (SWs) studied in the previous chapters. SWs have a wide sphere of applications, but here we shall consider only the following areas: plasma electronics, including devices utilized for generation of super high frequency radiation and plasma-antenna systems; plasma production, namely the plasma sources based on SWs propagation; surface plasmon-polaritons propagating in THz frequency bands; plasmonics devices, including bio-sensors, photovoltaic cells; SWs propagation in fusion plasma devices and plasma nano-science.

## 6.1 Surface Wave Applications for Plasma Electronics

Plasma electronics studies the generation of microwaves due to interaction between charged particle beams and eigen plasma waves. To create a plasma electronics device one can often introduce a plasma into an already established vacuum microwave device. This allows one to enhance the performance of the device due to: operation at higher injected beam currents and improvement of the control on the wave-beam interaction region. In its turn it leads to higher output power, higher efficiency and ability of operating frequency tuning. These effects have been observed [1] in the following electronic devices: plasma-filled relativistic gyrotron, plasma-filled smooth waveguide amplifier and relativistic plasma-filled backward wave oscillator (BWO). It should be indicated that in a plasma-filled smooth wall waveguide, excitation of slow waves by electron beam injection becomes possible. In paper [1] the authors studied the case of a relativistic BWO, applying a strong and a weak guiding magnetic field $0.2 \div 1\ T$, when the peak plasma density was higher than $10^{12}$ cm$^{-3}$. Their experimental work demonstrated a significant increase in operating efficiency and a tunable up shift of the operating frequency of the plasma-filled device as compared with the case of a vacuum device.

V. Girka et al., *Surface Flute Waves in Plasmas*, Springer Series on Atomic, Optical, and Plasma Physics 79, DOI: 10.1007/978-3-319-02027-3_6,

In the paper [2], results of theoretical and experimental studies of a large diameter microwave oscillator, which operates in the Giga-Watt power range, are presented. The device consists of a large cross-section, slow-wave structure with a unique profile of wall radius specifically designed to support surface waves (SWs) and to provide a strong beam-wave coupling at moderate voltage (500 kV), an internal adjustable microwave reflector, a coaxial microwave extraction section, and co-axial magnetically insulated field emission electron gun. The authors reported on the results of preliminary experiments carried out at 8.3 GHz, where the power level exceeded 0.5 GW and an efficiency of 15 % has been measured [2]. They have found that in waveguides with a large diameter SWs must be supported by specific wall structures in order to be synchronous with the electron beam. The features of SWs are crucial in order to avoid mode competition, and the experiments showed that a large coupling impedance is important to obtain high efficiency interaction. A periodic structure, which satisfies the requested conditions, was found by the authors. Their experiments allowed them to conclude that increasing the beam current to 16.8 kA had a small effect on the radiated power, but resulted in a down shift of the operating frequency by MHz.

The authors summarized that single-mode operation of a large-diameter SW generator has been successfully demonstrated, that the application of a moderate voltage electron beam can generate Giga-Watt radiation, and that additional optimization of the extraction section (coaxial and output window regions) is required in order to increase the radiated power.

One of the main purposes of plasma electronics is to develop devices, which will be able to generate electromagnetic radiation in the high frequency range, which could be powerful, tuneable and small sized. Reference [3] is devoted to the problems of elaboration of different schemes of microwave generators for super-radiation of extended electron bunches, which move above corrugated metal plates. One of these mechanisms is excitation of SWs by the electron beam in a corrugated structure. For the electron current 1,200 A/cm, corrugation period and depth $15 \times 10^{-3}$ cm and $5 \times 10^{-3}$ cm, there it is estimated that one can obtain electromagnetic pulses with the power $\sim 5$ MW and duration $\sim 40$ ps. Thus one of the main conclusions of the report is that THz band super-radiance can be achieved by utilization of the Cherenkov-type beam-surface wave interaction in corrugated waveguide circuits.

The same team of authors proposes in [4] to apply a planar relativistic SW oscillator with two dimensional periodic gratings for additional synchronizing electromagnetic emission from a wide sheet rectilinear electron beam in order to develop a Cherenkov millimeter wave generator with Giga-Watt output power. This paper belongs to the branch of plasma-electronic research, which study different mechanisms of Cherenkov emission when a sheet electron beam propagates in the vicinity of a periodically corrugated surface. In [4], a quasi-optical approach proposed by the authors in their earlier papers for describing the formation of a SW located nearby a shallow corrugated metal surface, is applied for the case of two-dimensional feedback. Moreover the authors of [4–6] also discuss the possibility to realize a cylindrical version of a two-dimensional SW oscillator driven by an annular electron beam. This idea is very interesting just for us because the materials presented in the previous Chapters of the book are devoted just to SWs propagating in waveguides with cylindrical geometry. Thus experimental studying of their interaction with an annual electron beam seems to be useful for demonstration of practical application of the surface flute modes.

The paper [7] is devoted to review accomplishments in the theoretical explanations and experimental performances of plasma–assisted slow-wave oscillators (pasotrons). This type of plasma electronic devices distinguishes from other type devices by two special features, namely: it has a plasma cathode electron gun to generate a high-perveance electron beam and it has no guiding magnetic field. Its slow-wave structure can be either a corrugated waveguide or a helix. Thus the electron beam interacts with by both axial and radial electric fields. The eigen electromagnetic waves observed there have their maximum of amplitudes nearby the walls of the pasotron; this is a specific peculiarity of SWs. That is why two-dimensional motion of the employed electron beam in this case allows one to enhance the efficiency of their interaction unlike in the case of application of an axial external magnetic field, which makes the electron motion like a one-dimensional one. Experiments conducted at that time allowed the authors to report on obtaining 50 % interaction efficiency and 1 MW of radiated power at an operating frequency $\sim 1.2$ GHz. In future the main efforts of the authors will be directed on understanding the physics of ion focusing of the electron beam and the interaction between the beam and the eigen fields of these slow waves. Thus they hope to achieve a microwave power of $\sim 2$ MW.

Cherenkov type interaction in corrugated waveguide between slow waves and an electron beam is studied in [8]. Because of the spatial distribution of the studied SW field one must direct the beam close to the waveguide wall. Thus the authors have applied a strong magnetic field $B_0 > 2T$, which was created by a superconducting magnetic coil with cryogenic cooling in the case of high repetition rate of the pulses. The performed experiments [8] have shown that the output power was more than 1 GW, with an electron efficiency close to 35 %. The authors proposed the design of a backward oscillator and an orotron with wavelengths: 3 cm and 8 mm, respectively as prospective devices for microwave generation.

The paper [9] presents results of experimental investigation an operation of plasma Cherenkov maser amplifier. The applied experimental apparatus consisted of four main parts: coaxial mode transformer, annual plasma column, plasma source, and transmission-line output. An external longitudinal pulsed magnetic field could be varied up to $B_0 = 2T$. The microwave source was a magnetron with the following parameters: radiation wavelength $\lambda = 2.42\,\text{cm}$, pulse duration $10 \div 100\,\mu\text{s}$, and output power $P = 50 \div 150$ kW. It was found that the plasma column consisted of filaments of a varying density, which were rotating azimuthally. The number of azimuthal plasma filaments was usually not larger than ten. The authors have decided to excite azimuthally symmetric coaxial plasma waveguide modes. The best results concerning experimental transportation of high power microwave radiation down the coaxial waveguide of 30 cm length were obtained for the slow $TM_{01}$ plasma mode. In this case $P \approx 150$ kW, $B_0 \approx 0.92\,T$, plasma density $3 \times 10^{13}$ cm$^{-3}$ where the input frequency was four times less than the electron plasma frequency: $\omega_0 \cong \Omega_e/4$.

The application of a plasma as an element of electronic devices has a lot of advantages connected with the peculiarities of plasma- filled waveguides. But enhancement of powerful HF oscillations excited by an electron beam in plasma-filled slow-wave structures is still far away from complete understanding. The authors [10] presented results of numerical calculations of the influence of the evolutionary electrodynamics characteristics, which can change due to nonlinearity of the plasma, on excitation of slow waves under the regimes of amplification and generation. Since the limitation of beam-wave energy exchange is caused by trapping of the beam particles into the wave potential wells the authors have studied inherent nonlinear mechanisms, which can lead to self-maintaining of the resonance between the particles bunch velocity and the slow wave phase velocity. It is found that increasing of an effective dissipation coefficient leads to increasing of the energy value, which can be lost by trapped bunches from the one hand, but on the other side this leads to decreasing of the quantity of trapped particles. Thus competition of these two mechanisms could lead to determination of an optimum value of the effective dissipative coefficient, which will allow one to realize maximum efficiency of slow wave excitation. It is shown in [10] as well that some enhancement of the efficiency of slow wave excitation could be achieved under the regime of stochastic modulation of the plasma.

The paper [11] presents results of experimental and theoretical investigations on a relativistic BWO (or karcinotron), which operates at the eigen $TM_{11}$ rotating wave. Decompression and cyclotron selection permitted the authors to obtain stable generation of microwave pulses, which have no after pulse. Many requested features of the karcinotron can be obtained by application of higher interacting eigen modes, among them power output, pulse duration, and stability of microwave generation. Higher eigen modes can be employed by the way of one-mode generation, and reduction of unwanted conversion of the operating mode to parasitic modes on various waveguide non-uniformities to an admissible level. To do that the authors applied a circular waveguide with a weak single threaded helical corrugation. To remove the degeneration of the frequency spectrum of the eigen waves propagating in mutually opposite azimuthal directions they have used a specific direction of the focusing magnetic field which was correlated with the waveguide corrugation. The karcinotron operated in the following regime: accelerating voltage $U \approx 0.5$ MV, beam current 2 kA, pulse duration $\tau \approx 20$ ns. Analyzing the experimental data [11] the authors have concluded that application of higher eigen azimuthally rotating modes has allowed them to realize stability of operation, which became close to potential barrier, to improve the control on the regime of generation, to reduce the value of the focusing magnetic field, and to suppress generation of the after pulse, which makes utilization of karcinotrons prospective for radar systems.

The authors in [12] reviewed results of integrated experiments devoted to operation of autoresonance masers and relativistic gyrotrons. As it is known, operation of the gyrotron is based on relativistic effect of the particle velocity on the cyclotron frequency. Thus if the energy of electrons increases the value of their cyclotron frequencies decreases, which means that the radiation frequency also decreases. Secondly, if the frequency of sub-relativistic electrons will change then they will go out from the cyclotron resonance with the excited wave, therefore they will transfer a small part of their energy into the wave. There are two obvious methods to compensate these defects of sub-relativistic gyrotrons. The first possibility is connected with application of the Doppler effect if the electrons will interact with the wave propagating not across an external magnetic field, but along it. Then like in the case of free electron lasers (FELs) the radiation frequency will be significantly larger than the electron cyclotron frequency. The second possibility is connected with application of non-dipole interaction of the sub-relativistic electrons which excites waves at higher harmonics.

Above mentioned authors have elaborated a special magneto-optical system supplied with a high current direct-action accelerator and a special pumped kicker for the electron beam that allowed them to improve the quality of the beam for the cyclotron auto-resonance maser (CARM). The use of an electron beam with a small velocity spread has enhanced the CARM efficiency to the level of FELs. But the work intended on further improvement of the beam quality must be continued. Another serious problem, which has been solved by the authors, is connected with excitation of parasitic modes, especially near-cut-off waves and backward waves. To avoid this problem the authors of [12] have applied the lowest transverse mode of a cylindrical waveguide ($TE_{11}$). Thus the maximum power at the wavelength of $7.9 \pm 0.1$ mm amounted to $13 \pm 2$ MW, at an efficiency of $26 \pm 5\,\%$.

As the first step to develop high-harmonic relativistic gyrotrons the authors have conducted experiments with large orbit gyrotrons (LOGs). They designed two devices: LOG-1 and LOG-2. In their experiments, selective excitation of $TE_{11} \div TE_{51}$ modes has been observed. The maximum signal was registered at the third cyclotron harmonic with a wavelength of 6.1 mm at an output power of about 100 kW. The opportunity to increase the efficiency of both devices is discussed. Thus according to the authors' experience [12] the development of high-efficiency CARM and/or LOG seems to be easier than the relevant FEL.

Results of theoretical and experimental investigations of three types of gyro-devices, which apply thin helical beams of electrons encircling the axes of axial symmetrical magneto-active microwave systems, are presented in [13]. The possibility of efficient simultaneous generation of an auto resonant mode at fundamental harmonic and a gyrotron mode at the second electron cyclotron harmonic is proved both analytically and experimentally there. The mentioned types of cyclotron resonance masers unlike conventional gyro-devices apply mildly relativistic electron beams, which have essential values of both axial and azimuthal velocities.

The first device combines features of a CARM based on excitation of a traveling RF wave and a relativistic gyrotron with excitation of cyclotron harmonics near the cut-off frequency. The main advantage of the CARM is its high electron efficiency. However one should take into account that creation of a high quality gyrating electron beam is a very difficult technical problem, while operation of the

CARM is very sensitive to the quality of the distribution function of the applied gyrating electron beam. The advantages of the second component of the proposed device, namely the relativistic gyrotron, are its stable operation and selective excitation of a near-cutoff cyclotron mode in a simple cavity. But it should be mentioned that its efficiency decreases with an increase of the electron cyclotron harmonic number. The authors have studied excitation of the $TE_{11}$ and $TE_{21}$ modes of a cylindrical cavity, which were excited by an encircling electron beam at the fundamental (CARM) harmonic and second cyclotron (gyrotron) harmonic. They have found that the proposed design of the cyclotron resonance device ensures a high efficiency of generated microwave generation due to excitation of the so-called "CARM" mode by bunches of electrons, which is provided by the second so-called "Gyrotron" mode. Just stable excitation of the "Gyrotron" mode allows one to fix the frequency of the generated "CARM" mode. The experiments [13] produced microwave radiation at the frequency 40 GHz with an electron efficiency of 22 % in the case of 460 kV electron beam voltage, 60 A beam current and 6 MWoutput power. The axial magnetic field, which was applied in these experiments, was about 1 T.

The second design is a large orbit gyrotron (LOG). The advantage of its operation is connected with the fact that the excitation of high cyclotron harmonics is stronger at relatively high energy of the beam electrons. In this device the authors [13] have once more applied a thin mildly relativistic electron beam encircling the axis of the utilized electro-dynamic system. The beam excited only co-rotating waves, which have azimuthal mode numbers equal to the resonant harmonic numbers, respectively. Selective excitation of the first five cyclotron harmonics ($TE_{11} \div TE_{51}$ operating modes) was demonstrated in prototype experiments with a short-pulse LOG. The obtained results were applied to develop a long-pulse LOG with a thermionic gun. To produce an electron beam, which would be able to gyrate around the axis of the wave-guiding system with a minimum displacement of its guiding center, with a perpendicular to parallel velocity ratio of 1.2–1.5 in a resonant magnetic field $1.6 \div 1.7$ T, the authors have designed a two-stage electro-optical system. In the first stage a thin electron beam is formed, and in the second one it acquires the requested transverse velocity passing through a kicker magnetic field. Measurements have shown that the efficiency of microwave generation was about 3.5 % at an output power of $70 - 80$ kW.

Interaction between a gyrating electron beam and eigen modes of a cylindrical waveguide has good prospects to be applied for efficient, broadband amplification of microwave radiation [13]. But in the limiting case of small axial wave numbers the bandwidth is limited by the dispersion of the eigen waves and in addition by the danger to excite undesirable parasitic oscillations. On the other hand, in the limiting case of large axial wave numbers Doppler broadening of the resonance restricts the bandwidth. Unlike to conventional gyro-devices, which have serious difficulties connected with their high sensitivity to the quality of the electron beam and to excitation of spurious oscillations, the authors have proposed to supply a waveguide with helical corrugation. Experiments at 9.4 GHz with this new type of gyro-TWT (traveling-wave-tube) have shown that the output microwave signal had a peak of 1.1 MW which corresponds to 29 % efficiency. This allowed the authors to make the conclusion [13] that such a broadband gyro-TWT with efficiency close to that of a comparable conventional gyrotron has some prospective as efficient high power amplifier for applications in radar systems and broadcast telecommunications.

We consider that there are two possible mechanisms, which can be applied for surface flute waves (SFWs) excitation, namely by application of charged particle beams or external alternating electric fields. The last references discussed in this section belong to the group of excitation of bulk waveguides waves by gyrating electron beams. But as it is indicated at the beginning of the section plasma electronics is closely related to vacuum electronics. Thus the papers [11–13], especially their parts concerned application of electron beams gyrating along large Larmor orbits, which encircled an axis of the applied waveguide, can be considered as a definite confirmation of a possibility to utilize such type electron beams for excitation of SFWs considered in the Chap. 2. Thus if one can inject an electron beam into the gap between plasma column and metal wall of the waveguide (see, Fig. 2.1, where the region indicated as "dielectric" can be changed by "vacuum and gyrating electron beam"), then this beam will be able to excite eigen SFWs.

We should remind that the radial distribution of the SFW field is described by Bessel functions of the first and second kinds (like any bulk wave) just in the vacuum region, and only out of this region (namely in the region of the plasma column) their field is described by modified Bessel functions which is typical for surface waves. Moreover, a gaseous plasma column can be exchanged by a semi-conducted cylinder then (taking into account that the density of such uniform and stable plasma is larger than $10^{14}$ cm$^{-3}$) one can obtain electromagnetic radiation in the THz frequency range. By the aid of these references, we would like to emphasize a prospective direction of further investigation of SFWs.

## 6.2 Surface Wave Applications for Plasma-Antenna Systems

At relatively recent time surface waves (SWs) have obtained a new field of application, namely plasma antenna systems. This new class of antennas is based on utilization of a plasma as medium with sufficiently high conductivity instead of metal wires and/or metal bodies. Taking into account the possibility to be quickly switched-off and/or switched-on, one can consider this antenna first of all, for military application. But it can be applied as well for civilian needs, mainly for development of technologies for wireless communication and for special selective shielding reflectors see e.g. [14, 15].

One of the first serious physical papers devoted to problems of plasma antenna design, which was published in a well-acknowledged journal was article [16]. It contains a report on results of measurements of efficiencies up to 50 % and radiation patterns for a single cylindrical plasma antenna. The authors proved that the distribution of electric current along the plasma antenna can be controlled by the plasma density, therefore plasma columns can be used instead of metal elements in communications antennas. In their previous experiments, the plasma has been produced by electrodes located at opposite ends of a dielectric tube, which contains a plasma column such as in a mains-driven fluorescent tube. This technique has some disadvantages, the main of them are as follows: complexity of this construction and large possibilities to disturb the plasma antenna radiating properties, which can increase the radar cross section of the antenna. In paper [17], it is proposed to drive a non-magnetized plasma sustained in the discharge from one end of the discharge tube, by excitation of the axially symmetric SW. In this case there is no need to apply two electrodes, which is a great advantage of the proposed plasma antenna design. But at the regime of low plasma density, the SW becomes heavily damped and has a shorter wavelength than current oscillations on a metal wire antenna. Generally the authors concluded that the efficiency of the proposed plasma antenna design is high sufficiently for RF communications.

Experiments performed by the authors of [18] have confirmed that propagation of the SW, which sustains a plasma column, can replace a metal rod as the guiding element in RF antennas. Moreover such plasma antennas will be of low radar detectability and negligible mutual coupling in the de-energized state. Their experiments confirm that the plasma antenna efficiencies and their noise level will be approximately the same as compared with the case of a corresponding metal antenna. It is also shown that the relatively high efficiency of the SW driven plasma column can be predicted by a simple theoretical analysis.

In addition to such an important feature as a low radar cross-section, the proposed antenna system, which applies a plasma column as the conductor in a RF antenna will be reconfigurable with respect to shape, frequency, bandwidth and directivity. The authors concluded that cylindrical plasma discharge tubes can be applied as antenna array elements, whose weight, required space and cost are less than that for any metal construction. Using the idea of [17], the authors have elaborated plasma antennas for the frequency range $3 \div 300\,\text{MHz}$, which applied the excited SW. The technical advantages of this construction of the plasma antenna are as follows: no electrodes, which are sources of plasma contamination; simplification of the antenna design as compared with both dual electrode and laser driven sources; minimization of the antenna detectability by radar and completely elimination of the mutual coupling when the antenna is de-energized. Plasma antennas of such construction will be maintained by about one-fifth of the power required for an equivalent fluorescent lighting tube. The authors indicate that the measured efficiency of the proposed plasma antenna construction is in good agreement with theoretical calculations. Thus the SW-driven plasma antenna produces sufficiently low noise, which makes it prospective for application in the field of broadcast communications.

It should be mentioned that the scientific company under the leadership of T. Anderson, President and I. Alexeff, Chief Scientist has performed an intensive investigation in the area of plasma antennas design, construction and application. Scientific collaborators of this company reached indisputable achievements in this sphere of plasma science. In the paper [19] the authors have demonstrated results of their experiments with various plasma antenna systems. Their activity was focused to the following main directions:

1. *Expansion of the working frequency range*. Their antenna systems can operate at a wide frequency range (500 MHz–20 GHz), which makes them useful for many military purposes and wireless communication.
2. *Making the plasma antenna invisible*. It was shown that if the external electrical supply is switched-off then the discharge stops and the plasma antenna reverts to a state of an ordinary dielectric tube, which has a small radar scattering cross section and is practically invisible for radars.
3. *Protection from electronic warfare*. The authors have demonstrated that their antenna system can be so configured that electronic-warfare signals with high-frequency will pass through the antenna without appreciable interaction, while it will transmit or receive signals at a lower frequency.
4. *Operation in the regime of shielding*. Creation of high-density plasmas in the discharge tubes of the designed plasma antenna systems allows one to apply them as reflector. When it is placed over a receiving horn and is switch-on, it prevents an unwanted electromagnetic signal from entering. When the plasma antenna system is switched-off, the signal passes through freely.
5. *Improvement of mechanical robustness of the antenna system*. The authors have developed two kinds of robust plasma antennas. In the first design, glass tubes intended for plasma confinement are encapsulated in a dielectric block. In the second one, the plasma antennas are composed of flexible plastic tubes (but they can be cooled additionally).
6. *Reduction of electromagnetic noise of the antenna system*. It was found that the noise generated by a plasma antenna is not a difficult problem. In the present case the level of this noise is comparable to the noise of a copper wire antenna of the same configuration, but in future the authors plan to realize some additional techniques to reduce the noise level.
7. *Alternative design of plasma antenna system*. The authors have presented as well results concerning the elaboration of a coaxial plasma waveguide consisting of an outer metal shell and the plasma discharge tube, which plays the role of the inner conductor. This device can be applied as a switch. The advantages of the proposed construction are that the metal shell of such a waveguide does not transmit electromagnetic signals if the plasma is energized, and this system has a small cross section for radar signals.

As compared with the results published in the previous work the mentioned company has succeeded enhancing of the plasma antenna efficiency for transmission and reception of electromagnetic signals. In [20] they have concluded that plasma antennas can work now as well as metal antennas. Their results concerning plasma antennas, which were operated in pulsed regimes; demonstration of transmission and reception of signals in noncurrent-carrying plasmas; no nonlinear effects have been observed in high-power regimes so far. Experiments devoted to the problem of noise in plasma antennas have shown that in the most cases, noise was a few decibels higher than that for metal antennas, which was always used for reference. Interesting results were obtained for solving the problem of the ignition of the plasma antenna influenced by a pulsed RF field. The authors have come to the conclusion that the plasma may or may not ignite in an intense RF field, because first of all plasma start-up depends on the design of the plasma antenna.

Experiments performed in [21] are devoted to investigation of plasma antenna properties for the cases of different plasma antenna construction. The main method of the paper is testing various configurations of the plasma column, which presents the main element of the antenna system. In the initial configuration, the plasma antenna consisted of a 30 cm long plasma column, which was excited by SWs. But it is known that the efficiency of plasma antenna operation can be increased either by increasing the antenna length or by using antenna arrays. Doing this one can take into account that for construction of very long plasma antennas huge power supplies, very long glass tubes and definite additional technical design intended on enhancing robustness of the antenna system are required. Thus application of the antenna system, which contains a long plasma column, is a difficult engineering task. In this case a phased array plasma antenna seems to be the best alternative to increase the antenna efficiency. Certainly, the construction of an array plasma antenna by conventional methods is also a difficult task because it needs to calculate the complex geometry of the antenna elements and to sustain definite phase relations between these elements. However, for a linear array the authors [21] have found a relatively simple way to avoid these difficulties based on the possibility to maintain the gas discharge volume as collection of volumes with spatially periodical plasma densities.

It was found that by variation of the supplying operating parameters, namely, type and operating pressure of the working gas, value of the input power and its frequency, radius and length of glass tube, a single linear plasma antenna, which consists of single plasma column, can be transformed into multiple small plasma antenna elements. The authors have also found a possibility to control the number, length, and separation between two plasma antenna elements. The performed experiments allowed them to investigate the efficiency of the antenna operation, its radiation power patterns and directivity of the radiation. The obtained results confirmed the author's assumption that this configuration of plasma antenna can be treated as a phased array vertical linear plasma antenna, which produces a more directive radiation pattern as compared with a single plasma antenna [21]. Their study has revealed the advantages of a plasma antenna over the conventional metal antenna due to its re-configurability and directivity of its electromagnetic radiation.

The paper [22] is devoted to consideration of a novel reconfigurable antenna concept based on the dynamic excitation of metal-like conductive plasma channels in high-resistivity silicon, which are activated by the injection of a DC current. These plasma channels are precisely determined and formed by the aid of high-resolution technology developed for fabrication of silicon wafers. The proposed reconfigurable plasma antennas enable to realize regimes of frequency hopping, to perform beam shaping and to provide steering of the beams without utilization of a complex and high-cost external RF supply. The authors hope that this concept of plasma antenna will allow one to obtain a phased array just at reduced cost as compared with conventional antenna systems. However, they consider challenges such as the non-linear features of the P–I–N semiconductor junction and the needed increase of the antenna efficiency. Thus further investigations are required.

According to the authors' opinion their novel approach, which applies P–I–N semiconductors junction, requires much lower power levels, less expensive power sources and much more direct interface to the external feed supply as compared with optical techniques. The feasibility of the proposed antenna design is depended on the possibility of creating sufficiently conductive patterns on semiconductors. It has been demonstrated that carriers of electrical charge in semiconductors form a plasma, which causes the semiconductor to behave like a metallic medium. The authors believe that future improvements of the proposed design of the plasma antenna will prove its feasibility, adaptivity and re-configurability.

Paper [23] contains results of experimental and theoretical studies, which were devoted to the examination of plasma antenna excitation applying an argon discharge sustained by SW propagation. The operation parameters of the device were as follows: working frequency 500 MHz, RF power up to 120 W and working gas pressures $0.03 \div 0.5$ mbar. The experimental observations show that the plasma column length increases with the square root of the applied power and its density decreases approximately linearly from the wave launcher to the end of the plasma column. This confirms the conclusions of a global theoretical model. The noise generated by the plasma discharge was measured for the operating frequencies $10 \div 250$ MHz and its excess level was found to be $17.2 \pm 1.0$ dB above 290 K, which corresponds to an Ohmic thermal noise at $1.4 \pm 0.3$ eV. Theoretical and experimental studies of the plasma antenna radiation pattern allowed the authors to conclude that the linear variation of the conductivity and the finite value of the column resistance lead to an increase in the width of the main lobe.

According to the authors' opinion the proposed plasma antenna design has the following advantages over conventional metal antennas systems:

1. Taking into account that the plasma antenna is energized for a very short time when communication takes place one can conclude that it is suitable for military HF communications. Since the tube operates by SW propagation under the regime of single-ended excitation it will be difficult to be detected by hostile radar.
2. Since plasma antenna arrays can be quickly reconfigured without excitation of perturbations from switched-off plasma elements, switched-off plasma antennas will not couple unwanted high-power signals into nearby receivers. Thus they can be applied in cases, which require utilization of different antennas for several communication systems, which should be located in close proximity. Hence it simplifies, e.g. the layout of ship-borne maritime communication systems.
3. Since plasma elements of the designed antenna can be energized and de-energized in microseconds, fast transmission of data at a low-frequency carrier can be switched at the end of each bit, thus minimizing signal degradation by antenna ringing.

4. Since the effective length of the plasma antenna can be varied by the aid of the applied RF power, this design of plasma antenna allows one to perform its reconfiguration for different transmitting frequencies.

The conducted experiments have shown that if the discharge tube becomes completely filled with plasma then the SW reflects from the end of the tube and deposits more energy into the top part of the plasma. Thus one can obtain a more uniform distribution of the plasma along the column. High density of the discharge plasma is required in order to maximize the efficiency of the plasma antenna and to minimize the plasma noise. The authors have found that at lower frequencies the noise of plasma antennas increase rapidly. They explain this fact by possible oscillations of plasma ions.

It is known that ionospheric clutter is one of the main clutters degrading, which occurs during the performance of HF surface wave radar (HFSWR), thus its suppression is an important problem for improvement of the radar operation. In [24], the authors proposed a new orthogonal projection method, which allowed them to enhance the HFSWR capability for target detection sufficiently. Thus it becomes possible to detect vessels on the ocean surface confidently at distances up to a few hundred kilometers. Ionoshperic clutter occurs because not the whole part of the vertically polarized HF electromagnetic wave emitted by the HFSWR propagates along the ocean surface; some portion of power propagates upward and is reflected from the ionospheric zone and due to Doppler shifting and widening of this signal electronic interference occurs on the radar display.

The proposed method for echoes analysis allows one to avoid installation of any auxiliary electronic element in the antenna system and to provide sufficient enhancement of the target detection capability even in the case of strong ionospheric clutter [24]. All the obtained conclusions are confirmed experimentally.

Book [15] is devoted to problems of practical utilization of SW propagation in gaseous plasmas, which are sustained in dielectric discharge tubes. The attention is paid mainly to the following possible applications of the SWs: plasma antennas and transmission lines, plasma reflectors and lenses, and plasma radiation. The fifth Chapter of the book deals with operation of plasma filled microwave devices such as: filters and phase shifters. To obtain a plasma in a special dielectric tube the author proposes to ionize a neutral gas. As a rule it is one of chemically inert gases like helium, neon, argon, xenon and krypton; mercury vapor can be applied as well. To perform the ionization process one can either utilize a DC current flowing between two or more electrodes, which have to supply a discharge tube or apply SW propagation.

Using the simple model of a cold non-magnetized plasma it is shown that the creation of a special "critical surface", where the plasma density has such a large value that the frequency of the studied signal is less than the electron plasma frequency, allows one to obtain a medium with negative dielectric permittivity, which will not permit this signal to pass through this surface. This physical phenomenon is the base for plasma reflector operation. So it is clear that application of a weak ionization strength will allow one to decrease the plasma electron frequency so the signal will pass the plasma layer. But the plasma shape and its dielectric properties can be designed in such way that it will act as a lens system as well.

The main concept of plasma antennas consists in the application of coupling between eigen plasma waves and electromagnetic waves, which can be radiated from a tube filled with this plasma. In our case, the surface flute X-modes are leaking modes, thus there is no necessity to establish coupling between them and electromagnetic waves. At least flowing of a weak electric current will allow one to realize coupling of the flute X-mode with the flute O-modes, which are of bulk type and therefore can be easily radiated from the waveguide.

The idea of a plasma filter is based on the features of the plasma, which reflects or transmits electromagnetic signals depending on the relation between the signal frequency and the cut-off frequency of the plasma. So changing either plasma density (in the case of non-magnetized plasma) or the value of an external magnetic field (in the case of magnetized plasma) one can tune the operating frequency and change the frequency band of the device. Thus book [15] provides a possibility for readers to get acquainted with a wide spectrum of practical applications of plasma antennas, using a simple model of a cold non-magnetized plasma without deep consideration of physical mechanisms of the studied phenomena.

More complete information about the elaboration of plasma antennas is presented in book [14]. It presents much more information than Ref. [15]. It concerns not only the presentation of exact mathematical solutions for scattered and reflected electromagnetic fields, far-field radiation patterns and the detailed description of the performed experimental work. There are new (compared with Ref. [15]) chapters devoted to plasma antennas nesting, windowing and steering; such designs allow one to elaborate plasma antenna systems operating in multi-frequency bands and multi-radiation lobe regimes, to obtain zero mutual inductance for extinguished antennas, which cannot be obtained by application of P–I–N diodes in the case of conventional metal antennas.

A special chapter of Ref. [14] is devoted to discussion of the results published by Borg et al. [16] studying the problem of decreasing noise of plasma antennas. It is indicated there that to decrease the noise level one can apply a higher operating frequency in the GHz range and a regime of low pressure of the operating gas, which fills the discharge tube unlike the experimental conditions, which were realized in the experiments of [16]. Two new chapters are devoted to the elaboration of smart plasma antennas and satellite plasma antennas, which benefit from low thermal noise. It is mentioned that in the past smart and satellite antennas have been applied for military purposes, but now these designs of plasma antennas can be utilized for establishing GPS—free positioning, for reconfiguration of signal beam width and numbers of lobes of a transmitting signal, for enhancing the useful part of the detected signal from satellite due to application of focusing and steering of the microwave beams, etc. Thus book [14] gives the readers not only the ability to understand the present advantages of plasma antennas over conventional metallic ones, but also to assess the prospects for further development of plasma antenna systems.

The case of a plasma antenna modeled by a cylindrical dielectric tube filled with non-magnetized plasma is under consideration in [25]. The study is carried out numerically using the solution of Maxwell equations and the kinetic Boltzmann equation. It is concluded that for an electromagnetic signal with a frequency of about 500 MHz used for ionization of the neutral gas with a density of $10^{16}$ cm$^{-3}$ one can expect to obtain the best impedance matching of the plasma antenna. Increasing of the frequency leads to increasing of the antenna radiation efficiency.

In the past decade, considerable attention has been attracted to elaboration of different plasma-like meta-materials, which could be applied in plasma electronics, and extensive studies have been performed to design novel high-quality antennas using these meta-materials. In [26], the authors consider the possibility to apply meta-materials in order to shrink the size of patch antennas. Namely, a two-dimensional cavity formed by a double positive-epsilon negative bi-layer waveguide is considered and studied to evaluate the radiation ability of its corresponding patch antenna. Based on the cavity model analysis of patch antennas, it is shown that good efficiency for broadside radiation of such a cavity-based rectangular patch antenna can be achieved when one layer of the cavity is partially shielded by perfect electrical conductivity boundaries.

Paper [27] is devoted to some theoretical problems related to the radiation of plasma antennas. The authors have analyzed two kinds of mechanisms of plasma antenna radiation. The first one is the radiation of an electric current excited by SWs, which is similar to the radiation of a metal antenna. In the second case electromagnetic radiation is connected with excitation of some transmission modes. This mechanism is similar to the radiation of a dielectric antenna; it can be realized because the electromagnetic waves partly cannot satisfy the total reflection condition in the plasma antenna for some specific frequency bands. The radiation properties of plasma antennas were studied numerically. The design of the antenna is as follows: a cylindrical dielectric tube is filled with non-magnetized uniform plasma. According to the conclusion of the authors the provided results can be important for further experimental investigation for construction of effectively controlled plasma antennas and plasma antenna arrays.

Experimental studying an electric dipole surrounded with a magnetized plasma shell [28] has proved that the input impedance of such an antenna could be matched with a feeder in a wide frequency range near the lower hybrid resonance. Radiated power of the studied plasma-antenna system can be essentially increased compared with the case of a conventional metal dipole antenna. An antenna immersed into non-magnetized plasma has resonant properties at the operating angular frequency $\omega_0$, whose value is in the vicinity of the electron plasma frequency $\omega_0 \leq \Omega_e$. This is connected with vanishing of the antenna input reactance. In this case a surface wave can propagate along the plasma surface, and the plasma reactance has an inductive feature. But in the case of a non-magnetized plasma often the collisional frequency is about the operating one, because of only low plasma density can be realized in this case. Unlike in the case of magnetized plasma a negative value of the plasma permittivity can be obtained for operating frequencies below the lower hybrid resonance.

So authors of Ref. [28] reported on results of experimentally studying a magnetized plasma antenna system operating in the frequency range 30 kHz $< \omega_0 <$ 2 MHz. A metal dipole with $2l = 80$ cm immersed into a magnetized plasma has been located in a glass tube with a length of 100 cm and a diameter of 16 cm. The working gas was argon under $10^{-3}$ Torr pressure. The magnetic field of about $10^{-2}$ T was produced by two Helmholtz coils and the maximum plasma density was $10^{13}$ cm$^{-3}$. It relaxed with the characteristic time constant 500 μs. Experiments [28] have proved that such a magnetized plasma antenna system has matching properties for frequencies less than the lower hybrid resonance and that application of the magnetized plasma shell allowed one to increase the radiated power by almost three orders of magnitude compared with a convenient metal dipole antenna without plasma shell.

Therefore the experiments described in [28] can be considered as proves of the possibility to apply the generation of surface flute waves propagating around a cylindrical metal rod immersed into magnetized cold plasma for the development of plasma antenna systems.

## 6.3 Surface Waves Propagating in the Plasma Periphery of Fusion Devices

The review [29] is devoted to a wide range of problems of ion cyclotron resonance heating (ICRH) and other applications of IC waves in thermonuclear fusion plasma devices, which should be solved for future application of the IC frequency wave technologies in the International Thermonuclear Experimental Reactor (ITER) being built in Cadarache, France. The author has mentioned that whenever the RF waves pass the plasma periphery, their power can be lost through parasitic absorption like collisional damping, wall heating, non-resonant damping, etc. Using the theory of so-called coaxial modes [30], which propagate in the coaxial cavity formed by the toroidal chamber wall and the plasma column with the dispersion relation: $k_{II}^2 = \omega^2/c^2 - k_{\varphi}^2$, here $k_{\varphi}$ is poloidal wavenumber, the author has tried to answer the question: what is the minimum value of $k_{II}$ (parallel wave number) requested for optimal heating of ITER. Assuming that all power launched in coaxial modes does not contribute to bulk plasma heating, he estimated the lower limit of the parallel wave number to be 3 m$^{-1}$ for ITER. The importance of studying SW propagation during ICRH in fusion plasma devices was already emphasized earlier in [31–34]. The authors of [31–34] concluded that just propagation of different SWs is the most probable reason of such undesirable processes in fusion plasmas as plasma periphery heating and generation of impurity ions with large atomic numbers.

In the paper [35] it is proved that slow waves excited in JET and other tokamaks during ICRH regimes can produce larger momentum than fast waves. This allowed the authors to predict that these waves could be able to drive plasma flows. On the other hand there exist some contradictions between the theory, which indicates RF driven rotation and the preliminary experimental results. So both experimental and theoretical work will be continued.

Paper [36] can be considered as a mile-stone, which overviews the main results in solving problems of IC wave application in large-scale fusion devices like JET, which are supplied with superconducting coils for creation of the confining magnetic field. To support continuous periodical operation of ITER one has to solve the problem of efficient conditioning technique for the walls of the fusion device chamber, which would be able to reduce the impurity content in the chamber, to control the hydrogen isotopic ratio on the inner surface of the chamber walls, to mitigate the long-term tritium inventory build-up in the chamber and even to recover the chamber walls after some disruptions.

Since this conditioning will be performed under the condition of presence of the residual external constant toroidal magnetic field produced by the superconducting coils, glow discharge conditioning (GDC) of fusion devices, which was frequently utilized until now, will be useless in ITER. The authors propose the IC wall conditioning (ICWC) technique as an alternative to the GDC method. It is assumed that the technique will apply the ICRH antennas, which are already installed for RF plasma heating in present-day tokamaks and stellarators, working in different countries. Preliminary results obtained at JET and TEXTOR are encouraging and the authors estimate the ICWC technique as one of the most promising, which will be able to support ITER with routine inter-pulse or/and overnight conditioning. They see the following tasks for ICWC: first, to ignite safely and reliably an RF discharge by the ICRH hard ware, which already has been installed in the fusion chamber; second, to couple the main part of the RF power to low density ($\sim 10^{10} \div 10^{12}$ cm$^{-3}$) plasmas at the presence of large toroidal magnetic fields ($\sim 5$ T); third, to develop RF power absorption schemes intended for obtaining a more uniform plasma discharge and for enhancing the conditioning effect.

In our opinion the last task could be solved by application of discharges sustained by SWs. As it is known see e.g. [17, 37], gas discharges sustained by various types of SWs have a lot of advantages as compared with the case of discharges based on bulk wave propagation: SWs are slower than corresponding bulk waves, thus they can effectively interact with plasma particles; the radial distribution of SW fields has the peculiarity that the maximum values are located nearby the plasma interface, therefore just the application of SWs will allow one to obtain a more uniform plasma discharge; excitation of SW eigen modes is more effective as compared with the case of bulk wave excitation.

Paper [38] is devoted to study *Ni* impurity generation in the JET plasma during operation of two different types of ICRH antennas. In spite of the fact that additional ICRH always leads to impurity generation in fusion devices the authors state that the quantity of metallic impurities in JET plasmas does not significantly affect the plasma heating. They consider two possibilities to decrease the impurity concentration in the fusion reactor without changing hard ware: to shift the plasma away from the ICRH antennas or to increase the $D_2$ gas-puffing rate. The most interesting result obtained there is the possibility to reduce the *Ni* concentration by changing the location of $D_2$ gas injection.

It is know that a compulsory element of large fusion devices is a divertor intended for removing the helium ash produced by the reaction of fuel combustion in a thermonuclear reactor and various impurities generated in the magnetically confined plasma due to active interaction between the plasma and the first wall of the reactor vessel. Paper [39] is devoted to studying a flute instability in the divertor region of a tandem mirror machine. The GAMMA 10 A-divertor is proposed by the authors as a candidate for the upgrade of tandem mirror fusion device Gamma 10 at Tsukuba, Japan. The most dangerous instability in an open magnetic plasma system is a flute instability, which manifests itself by appearance of a small perturbation of the plasma volume on the surface of the confined plasma column, which moves azimuthally around the plasma column. The shape of the plasma column cross-section seems to deviate from a circular one. The quantity of such flutes located on the surface of the confined plasma column is described by the azimuthal mode number $m$ (see e.g. [40]). Then this small perturbation (let's consider the case $m = 1$) becomes larger and larger and at least this perturbation is touches the first wall of the reactor. This can lead to destroying the coating of the first wall and even to a failure of the plasma discharge.

The main purpose of the installation a divertor mirror cell is to carry on the simulation experiments of the divertor intended for application in a large torus like LHD, JET and ITER. Traditional tandem mirrors such as GAMMA 10 contain a space region with minimum value of an external magnetic field in order to stabilize the flute instability. But existence of such a region with minimum magnetic field enhances the neoclassical transport. Thus elaboration of a new type of divertor is intended for solving this difficult problem, namely to stabilize the flute instability without a strong enhancement of the neoclassical transport. The reduced MHD simulation performed by the authors allowed one to find an answer on the question: what is the influence of the divertor region in their new design on the flute instability? It is found that the divertor region plays a positive role in stabilizing bulk flute modes.

In their next paper [41] the authors continue investigations of the stability of bulk flute modes in the GAMMA 10 A divertor. They have carried out a stability analysis by application of three-dimensional treatment of the basic equations describing the fluctuation of bulk flute modes. It is found that the stability boundary of these modes is essentially dependent on the radial profiles of plasma density and temperature in the axisymmetric divertor mirror. They have calculated the linear growth rate of the instability of the bulk flute mode with azimuthal wave number $m = 1$ for various radial profiles of plasma density and temperature in the GAMMA 10 A-divertor. They found that the modes are stabilized by a special form of the applied magnetic field profile in an anchor mirror cell, while this stabilization is performed due to sufficient plasma compressibility. Changing the definition of the specific plasma volume of a magnetic field line applied in the bulk flute mode stability criterion did not lead to changing results compared with the previous definition of this specific parameter. Thus the results of paper [41] revealed that the linear phase of the flute instability is the same as those in the previous works on this topic.

The last two references are not devoted to the case of surface waves propagation but they concern the propagation of bulk flute modes and their stability. Thus they also confirm the importance of studying eigen surface flute modes, whose theory is presented in the previous chapters of book.

## 6.4 Surface Wave Application in Nano-Technologies

Review [42] is devoted to a wide spectrum of technologies for metal plasma production, mainly to energetic condensation of metal plasmas, and to the formation of nano-structures using metal plasmas. The plasma production techniques are based on application of ionizing metal vapour obtained in high frequency and strong current discharges. Special attention is paid by the author to different forms of ionized metal vapour deposition, namely magnetron sputtering, metal self-sputtering, high power impulse magnetron sputtering, cathode arcs and pulsed laser ablation. Metal plasmas produced by the indicated methods are already utilized in several industrial techniques, such as the deposition of ultra-thin layers in the computer and data storage industry and the coating of special tools. Examples of industrial spotless arc deposition for titanium with rates up to 1 μm/s have been realized.

Taking into account the possibility of surface flute X-modes to propagate along the metal interface, we hope that they could be useful to develop some technologies of solids processing, namely polishing and etching. Then by the aid of advanced diagnostics, deposition and processing techniques, metal plasmas can be used for fabrication of photonic and biomedical devices.

Production of nano-structures by metal plasma is studied there as well, including both the possibility to use the metal plasma for conformal coating or for completely filling trenches on a substrate surface with the nano-structures, which are primary prepared by lithography. The possibility to apply metal plasmas for fabrication of self-organized nano-composites (it means production of a matrix of amorphous material filled with nano-crystals) and nano-porous materials also is discussed. The initial material can be decomposed and then the deposition of the prepared metal plasma can be performed at the meta-stable phase and as a result one can obtain highly porous nano-structural materials. It is known [42] that metal plasma techniques allow one to obtain a trench filling and a conformal coating of a substrate surface down to the 100 nm size. Application of metal plasmas already allows one to fabricate super hard (H > 40 GPa) and tough nano-structures. For instance, high hardness of multilayer composite films can be obtained by cathode arc plasma deposition of titanium–aluminium and titanium–zirconium. This is known more than 20 years. But more interesting and prospective results can be achieved by the incorporation of ions of such unique metals as yttrium or vanadium to stable phases of the both *TiN* and *AlN*. During their further thermal treatment they can improve the oxidation resistance and also sufficiently reduce the coefficient of friction of the nano-structures (about twice in the high temperature range: $850 \div 950\,C^0$).

Let us remind that the term surface localized plasmon (SLP) [43] means collective oscillations of conductivity electrons in some metal nano-structure excited by electromagnetic waves, whose wave length is belonging to the visible light spectrum. Since the light wavelength is larger than a nano-structure scale the conductivity electrons, affected by it, oscillate nearby the nano-structure. In the paper [44], which can be considered as supplement to the US Patent Application [45], the authors present a solution of the problem: how surface localized plasmons can be used for enhancement of forward light scattering into a photoactive layer without over-shadowing by strong absorption of the light at the plasmon resonance wavelength of the *Ag* nano-particles incorporated into solar cells. It is shown that in this case, SLP resonances can be shifted to the low-frequency edge of the light spectrum, by varying the dielectric properties (therefore both the refractive and reflective indexes change their values) of this solar sell.

The authors have found that nano-particles located at the front surface of the solar cell perform trapping of the light at wave lengths, which are above the surface plasmon resonance. Nano-particles located on the rear of the cell provide trapping of the rest portion of the incident light. So they came to the conclusion that using nano-composite technique, namely introducing nano-crystallites of *Ag* into the commonly utilized silicon solar cells will allow one to produce large-scale photovoltaic devices.

Review [46] is devoted to one more practical application of surface plasmon resonance (SPR). It is shown that the SPR as an optical technique can be applied for investigation of biological interactions. Unlike most of the known techniques the SPR can analyze the dynamics of surface events in real time and, therefore using it one can determine and estimate rates of adsorption and desorption for a range of surface interactions. The authors conclude that utilization of the SPR to probe surface interactions has a lot of advantages as compared with other classical methods: it is able to rapidly monitor any dynamical process, such as adsorption or degradation; there is no necessity to perform difficult complex preparation of biological and/or medical samples and relevant tested medical reagents and/or reactives for the respective investigations.

It is shown there that the SPR is a prospective technique for studying processes of drug release from biodegradable systems and is a fast, simple technique to examine the response of biological fluids to biomaterial surfaces. Moreover according to the author's opinion the future development of application of SPRs is connected with elaboration of biosensors [see example of the production of a commercial biosensor (Pharmacia Biosensor, "BiaCore")] for the investigation of specific biological interactions including adsorption and desorption kinetics, antigen–antibody binding, and epitope mapping.

Paper [47] presents results of experimental investigations devoted to finding ways for widening the sphere of SPR applications as biosensors. It is confirmed that the SPR, which is exploited commercially in instruments such as the

"BiaCore", is one of the most effective methods of detecting protein molecules (especially enzymes and antibodies) motion relative to a metal surface (as a rule this is a golden or silver film). The authors have performed experiments intended to detect the effect of electric fields on these molecules and found that the application of an external electric field leads to attraction or reflection of protein molecules and colloidal particles on the golden film. The direction of the motion is depended on the molecules charge state.

Progress in elaboration of the first completely autonomous (electric power source is an ordinary 9 V battery) hand-held biosensor based on the surface plasmon resonance (SPR) technique, which does not require a computer interface for its operation is reported in [48]. Testing experiments have shown that its responsivity is as small as $3.3 \times 10^{-6}$ refractive index units. Thus the sensor can detect such biological toxins like ricin at a concentration of $200 \times 10^{-9}$ g/ml. The development of such type of sensors is connected with on-going needs of the security sphere to counteract the threat of bio-terrorism. So fighting against the threat of applying bio-toxins and dangerous organisms such as ricin and anthrax, we need systems that can quickly and correctly operate in situ. Concerning other possible applications of the elaborated bio-sensor based on the SPR, the authors indicate the following fields: diagnostics of animal and plant pathogens in order to organize a quarantine zone, gene analysis, remote rapid testing of a malarial strain, and water purity analysis. At least it can be applied as detector of unexpected chemical processes in laboratory devices.

Article [49] is devoted to investigation of electromagnetic band gap (EBG) structures, which are also known as photonic band gap materials (photonic crystals). Such meta-materials [50] represent periodic structures, which create forbidden frequency bands for propagation of definite electromagnetic waves. The periodicity of the structure is approximately one half-wavelength. The possibility of building up such materials with the requested electromagnetic properties makes them attractive for applications such as antenna systems [19], elements of millimeter wave electronic devices [49] and optical communication devices [51]. It is known that the woodpile structures are among the most popular EBG structures [52] that are explained by their following advantages: large bandwidth, robustness and relative simple fabrication.

In the papers [53, 54], it is shown that the woodpile structure can be used as antenna substrate, placing a planar antenna on top of its surface. Since no electromagnetic radiation can propagate within its band-gap the woodpile structure can act as a reflector for the microwave antenna radiation. Moreover utilization of woodpile structures for building up waveguides, whose width increases near one of the edges of the waveguide, allows one to elaborate horn like antennas [55].

The experiments conducted by above authors have shown that in the case of various truncated woodpile structures, additional surface modes appear, whose eigen frequencies could be located in the band gaps of the structures, which will lead to their coupling and interfering with the usual working frequency range of the EBG structure.

The authors of paper [56] have presented results of temporal characteristics measurements of surface-plasmon polaritons (SPPs) at terahertz (THz) frequency band. SPPs have been generated by coupling THz radiation into dielectric films deposited on flat gold surfaces. This allowed the authors to conduct measurements of thin dielectric film features and to demonstrate the capabilities of the application of SPPs for thin film spectroscopy.

SPP technique has confirmed its possibilities to measure optical and dielectric constants, dielectric film thickness, uniformity of dielectric coatings of metal surfaces with high precision and to control light arrays over length scales, which are much smaller than the light wavelength. This permits to apply SPPs in the area of plasmonics e.g., for fabrication of nano-scale photonic circuits. Other branches for application of this technique are medical diagnostics and three dimensional imaging systems for monitoring industrial processes. It is found that biologically active substances like proteins, ribonucleic and deoxyribonucleic acids have resonances in the THz range, thus SPP spectroscopy at THz frequencies could be applied for creation of new biosensors. Due to analysis of the obtained experimental results the authors point out the high sensitivity of this technique and the possibility to apply it for sensing of biomolecules with resonant transitions at THz frequencies.

Accordingly to opinion of the authors of paper [57] the SPPs, which propagate in the visible and infrared spectral ranges, are studied well enough, which is confirmed by their wide applications in spectroscopy and sensor development. However, the properties of SPPs in the THz range are far from complete understanding. Using a free electron laser with averaged power 10–50 W, wavelength $130-140\,\mu m$ the authors have focused the radiation on the input mouth of a parallel plate waveguide, which was formed by the surface of a sample and the lower face of a glass prism (both surfaces were covered with a $1\,\mu m$ gold layer). They have found that the features of SPPs are strongly dependent on the thickness of the dielectric layer deposited on the conductor surface. After reaching the end of the sample, SPPs radiate a strongly directional THz beam into the free space, which can be used for THz probing.

The paper [58] is devoted to studying the possibility to use modern metamaterials for an efficient guiding of surface electromagnetic waves in the Giga- and Terahertz frequency ranges. The authors have considered two models: first, surface wave propagation along the interface between a semi-bounded meta-material and vacuum, and second, surface waves guided by layers of metamaterials. It is found that surface waves can propagate as symmetric and antisymmetric modes in meta-material layer. Varying the layer thickness and some parameters of the material one can realize regimes of forward and backward propagation of the SWs. The authors have concluded that one can control the slowing down of the SWs.

Taking into account that surface flute waves can propagate either along a magnetized or a non-magnetized plasma interface we estimate their application for nano-technologies as a very prospective.

## 6.5 Surface Wave Application for Plasma Production

Plasmas obtained in different types of discharges are finding increasing use in materials processing, plasma chemistry, chemical analysis, and other fields. One of the methods to sustain gas discharges [59] is connected with excitation of electromagnetic surface waves (SWs). This paper reviews the physical principles of operation and the design of surface wave plasma sources. Since the wave launcher is the central component of any plasma source, this review presents a description of several compact, efficient and easy to operate launchers specifically intended for plasma generation. Using them it is possible to sustain plasmas in the frequency range from 1 MHz up to 10 GHz, at working gas pressures from $10^{-5}$ Torr up to few times atmospheric pressure. But in order to simplify the presentation, the authors have restricted their considerations to the case of a low pressure discharge sustained by SWs propagating along the axis of symmetry of a cylindrical dielectric tube. The length of the active zone of the discharge, where the HF power transfers to the plasma particles, is considered to be larger than the tubes' diameter.

It is proved experimentally that SWs, as eigen modes of the discharge tube, can be excited by a very compact wave launcher, which takes up only a small area of the discharge tube, the HF power being carried to the rest of the plasma column due to the discharge by the propagating wave. It was indicated that the obtained plasma is stable and reproducible because a stable mode of a definite SW can propagate there. Thus the level of electron density fluctuations was low enough. The authors underlined an essential difference between the case, when the SW propagates along the interface of a plasma column created by some other methods and the case, when the eigen SW sustains the plasma column by itself. The fact that the discharge and the SW, which sustains it, depends upon each other has allowed the authors to formulate an equation of the power balance between the power lost by the electrons and the power acquired from the SW field.

The papers [59, 60] have studied specific features of the process of SWs excitation for the cases of azimuthally symmetric (azimuthal mode number $m = 0$) and azimuthally non-symmetric ($m \neq 0$) perturbations. Summarizing the results of the experimental investigations obtained for the cases $m = 0$ and $m = 1$ they concluded that the $m = 1$ non-symmetric mode can be launched to sustain a plasma column only if the product of the SW frequency $f$ and tube radius $R$ exceeds a definite critical value, in the case of paper [60] it was 2 GHz cm. However, the azimuthally symmetric $m = 0$ mode can propagate and sustain the plasma for each value of the SW frequency, which results in the product $f \cdot R < 2$ GHz cm. This circumstance has led the authors of review [59] to develop various $m = 0$ SW plasma sources. Let us consider the main types of them.

The first of them is the LC Ro-box, which operates in the lower domain of the indicated frequency range (approximately $1 \div 100$ MHz). It consists of two separate units: the field applicator (or field shaping) module and a matchbox, whose impedance matching network contains LC elements. The second one is the stub Ro-box. This SW launcher uses the same field applicator as those of the LC

Ro-box, but its power feed and impedance matching are different. The HF power is supplied to it via a coaxial feeder through a capacitive coupler. This impedance tuner is constructed of semi-rigid cable, which penetrates radially into the field applicator module. The third launcher is entitled as surfatron. It was the first SW launcher, which has found a wide range of applications as power source intended for production of long plasma columns utilized for various physical studies, plasma chemistry, solids processing, laser engineering, etc. The surfatron is an integrated wave launcher, which performs both field shaping and impedance matching (the external stub used in the stub Ro-box launcher is now part of the surfatron body). The fourth type of SW launcher is the surfaguide. It has been elaborated because of restrictions in the power handling capability of the surfatron. Without doubt the surfaguide is the simplest of the surface wave launchers among those considered here. It consists of two parts: RF field applicator, which is a section of rectangular waveguide of reduced height, tapered symmetrically with respect to the launching aperture and a movable plunger, which is located at one end of the applicator. The last type of SW launcher developed for sustaining gas discharges by propagation of azimuthally symmetric SWs is the waveguide-surfatron. It is a surface wave launcher consisting of both waveguide and coaxial line elements. Here the microwave power is supplied to a rectangular waveguide section by a generator and an air insulated coaxial line section is attached perpendicularly to the wide wall of the waveguide, so that its inner conductor extends into the waveguide.

The results of experiments reviewed by the authors enabled them to conclude that applications of SW sustained discharges employ a range of operating conditions and plasma parameters available with gas discharges. Namely, these types of discharges allowed one to obtain over dense plasmas in relatively large discharge chambers. The plasma sustained by SWs propagation in such discharges is characterized by high level of reproducibility, convenience to utilize it for various technological needs and by low level of parasitic electromagnetic oscillations. This is explained by such useful features of the SWs like slow phase velocity, mono-mode regime of wave propagation and weak dependence of the wave impedance on plasma parameters [59].

According to the authors' opinion, the development of suitable wave launchers has been the main challenge of SW plasma applications, because the other elements of the plasma source can be elaborated by good established methods of electronic engineering.

Paper [61] is a one of the first reviewing articles devoted to the development of large-size, high-density $n_e \geq 10^{11}$ cm$^{-3}$ microwave plasma production. It concerns plasma production, which can be performed at low pressures (<20 m Torr) of the operating gas, without utilization of an external magnetic field. Deep understanding of the discharge physics has allowed the authors to elaborate a microwave plasma production technique, which is characterized by a stable operating regime, generation of a large-diameter plasma column (diameter is larger than the axial length of the column) sustained by azimuthally non-symmetric SWs, which were clearly observed in experiments using optical emission and microwave field measurements. The authors proved that just excitation of SWs gave them the possibility to reach such good results, thus after improvement of the antenna structure this technique will be commercially available for processing of large scale flat panel displays and solar cells with diameter larger than 1 m.

Comparative analysis of different types (including electron cyclotron resonance plasma, helicon-wave excited plasma and inductively coupled plasma) of plasma production methods has shown the advantages of this SW plasma source in comparison with other high-density sources. All that allowed the authors to conclude that in the regime of low gas pressure the slot-excited azimuthally non-symmetric SWs are attractive for sustaining a high density, large scale gas discharge, which produces non-magnetized plasma for solids processing technologies. So the theory of surface flute modes considered in the previous chapters of this book can be directly applied for explanation of the experimental results of Ref. [61].

Microwave production of plasmas is widely used in the fields of semiconductors production and solids treatment. It is characterized by the following features: higher electron density compared with RF capacitive discharges, possibility to operate at extremely reduced gas pressures, low energy of the ions striking the substrate, and a possible independent biasing of the substrate. That is why the authors of [62] propose to concentrate the main efforts on theoretical and experimental studying of SW produced plasmas influenced by an external magnetic field. It is found that in a magneto-active plasma column located in a dielectric tube of a finite thickness, fundamental surface hybrid waves $HE_{m,n}$ with azimuthal mode number $m$ and radial mode number $n$ ($n - 1$ is the number of zeros of the axial electric field $E_z$ variation along radial direction) can effectively sustain a microwave gas discharge. The produced plasma density increases with the power deposited into the discharge per unit axial length. Experiments prove that azimuthally non-symmetric SWs with azimuthal mode numbers $m = \pm 1$ have a larger attenuation coefficient than for the case of azimuthally symmetric surface wave. This fact can be considered as a witness on prospective application of surface flute waves for sustaining microwave gas discharges.

Results of experimental approbation of a new method of generation of a stable surface microwave discharge in a boundary layer near a body streamlined of supersonic airflow are presented in [63]. Breakdown characteristics, spatial distribution of electric field, dynamics, space-temporary evolution of gas, and other characteristics of this type discharge are measured. This investigation is initiated by the needs of supersonic aerodynamics influence the characteristics of gas flow near a surface of flying bodies. Since electrode discharges in such air flow are unstable, spatially non-uniform, and non-reproducible, the authors have paid attention to a new version of the electrode-less microwave discharge. Their experiments have shown that the most suitable type of discharge is a microwave discharge sustained by SWs. The SW can propagate in a gas as long as its energy is enough to create a plasma, whose density is no less than the critical density. Thus there is a self-sustaining system. The experimental set up consisted of a vacuum chamber, magnetron with pulsed power $P \leq 300\,\text{kW}$and wavelength $\lambda = 2.4$ cm, pulse duration $\tau = 1 \div 100\,\mu\text{s}$, system for microwave power input, system for creation of supersonic airflow with Mach number $M = 2$, and diagnostic system.

It is found in [63] that the minimum power, which is needed for generation of the discharge, decreases with increasing pressure in the air pressure range from 0.001 Torr to 0.1 Torr, then the minimum power increases with increasing pressure in the higher pressure range 10 Torr $< p <$ 1,000 Torr. The ionization frequency is an increasing function of the electron temperature, or (in other words) ratio of the plasma density divided by the electric field strength. The longitudinal size of the discharge was found to be a linear function of the pulse duration. According to the obtained results of a comprehensive program of investigations, the authors concluded that microwave discharges sustained by surface wave propagation provide stable uniform luminous plasmas, which coat the whole surface of dielectric bodies immersed into the supersonic airflow and that the flow with the Mach number mentioned above does not influence the characteristics of this discharge.

The paper [64] presents material, which is continuation of the previous work of these authors. They have found that at the regime of low air pressure $p < 1$ Torr, where the electron-neutral molecule collisional frequency is much lower than the eigen frequency of the applied surface wave, the discharge is very uniform and occupies a large volume, whose size increases with decreasing air pressure. The authors propose to apply such type of discharge for treatment of solid surfaces, etching, deposition of films and special coatings. Under the regime of high pressure 30 Torr $< p$ the discharge has a complex form of thin plasma channels (filamentations), whose diameter is less than 1 mm. Formation of such discharges gives rise to shock waves, thus it can be applied for ignition of liquid hydrocarbon fuel.

The paper [65] is devoted to investigation of fundamental problems of microwave gas discharges: low threshold of the discharges at a metal-dielectric interface; thermal-ionization instability; plasma resonance on its non-uniform density profile. Results of experiments for studying plasma generated by three new types of plasmatrons are discussed there.

The paper [66] presents a self-consistent model of a TM azimuthally symmetric SW discharge in hydrogen applying cylindrical geometry. The basic set of equations is derived from Maxwell equations in kinetic approximation. Conducted simulation and analysis of experimental data allowed the authors to make the following conclusions: the electron density in this discharge sustained by SWs exceeds the critical value for the 3 cm wavelength band and reaches $10^{13}$cm$^{-3}$, the velocity of the SW front is essentially dependent on the plasma resonance at the front of the discharge, a significant role in the kinetic of the discharge is played by ion-molecular reactions, which lead to ion conversion, and that the considered discharge is a very efficient source of hydrogen.

It is known that neutron beam injection will be applied for plasma heating and current drive in ITER. The basis of the injection method is neutralization of negative hydrogen ions accelerated up to MeV energies. The efficiency of neutralization determines the total efficiency of the neutral beam injection. Comparison of the gas neutralization method, which is presently used and the plasma neutralization method, which is developed by Skovoroda and Zhyl'tsov in [67] show an undisputed advantage of plasma neutralization for the conditions of ITER, namely under the conditions of a large volume of confined plasma and a large value of the applied external magnetic field. The concept of the device elaborated by the authors is based on application of a multi pole magnetic trap with the following properties: closed drift of electrons connected with the gradient of an external magnetic field, large magnetic field and large volume of the plasma neutralizer. The estimations made by the authors allowed them to affirm that the proposed design of plasma neutralizer will be more efficient compared with the usual gas neutralizer.

Taking into account that microwave discharges are intensively used in different applications, the authors [68] presented two examples of their applications. The first one is the plasma neutralizer, which can be considered as an attractive upgrade for Neutral Beam Injection Systems that will be utilized in ITER. The second one is the plasma separator for treatment of spent nuclear fuel and radioactive wastes.

The authors have obtained a sufficient increase of the neutralizer efficiency when changing the gas reagent by a plasma produced in a large magnetic trap under the regime of an electron cyclotron resonance (ECR) microwave gas discharge. This paper develops ideas proposed in [62] on advantages of ECR microwave discharges over other types of discharges and is a sequel of the previous paper [67]. It is reported that at present time utilization of a plasma neutralizer allows one to increase the neutralization efficiency up to 80 % in comparison to 60 % achieved in the case of a gas neutralizer.

Application of plasmas produced by ECR plasma sources for ionization of spent nuclear fuel components is the first step into a new technology developed by the authors. The separated elements are selectively heated in an ion cyclotron resonance chamber and finally selected in a toroidal magnetic field. The first experimental results allowed the authors to estimate the proposed design of a plasma separator for treatment of spent nuclear fuel and radioactive wastes as very good prospective [68].

Discharges sustained by SWs are under consideration of many groups of scientists because of their technological applications and also due to their utilization for fundamental investigations in various branches of physics. Concerning just the last indicated aspect the steady state of SW produced discharges especially at low pressures (less than 5 Torr) and relatively low frequencies (less than 5 GHz) is studied actively at present time. But on the other hand, there are many applications where long uniform plasma column can be generated for short living-time. According to the opinion of the authors [69] the maximum frequency, at which one can obtain a short time discharge, does not exceed 10 GHz at present time. In this frequency band the SW is usually launched into the discharge system by a surfaguide [59]. But its operating power is limited by gas breakdown in the region, where the SW is excited.

That is why the authors of the paper [69] have developed two novel types of launchers, which allow one to apply much higher value of microwave power for generation of SW sustained discharge under the pulsed regime. The first launcher is called as E-plane junction; it allows one to generate a long plasma column in a quartz tube placed in a single-mode waveguide. The tube had an inside diameter of 4 mm, an outside diameter of 6 mm, its length was 50 cm, and it was filled by various gases: air, argon etc., under the pressure $1 \div 60$ Torr. The experiments were conducted using microwave radiation with 3 cm wavelength, $\tau = 1 \div 15\,\mu s$ pulse duration, $30 \div 150$ kW pulse power, and $0.1 \div 1$ Hz repetition frequency. High degree of absorption of the input power (its typical value is $70 \div 90\,\%$) was observed over a wide range of gas pressures. That is why the authors have concluded that this type of microwave SW launcher allows one to extend the operating power level.

The second type of launcher applied the circular waveguide $E_{01}$ mode for excitation of the operating SW. The discharges were sustained in quartz tubes with $6 \div 9$ mm outside diameter, and 30 cm length. These experiments were carried out for microwave radiation with 8 mm wavelength, $\tau = 0.5 \div 10\,\mu s$ pulse duration, $30 \div 60$ kW pulse power, and $10 \div 10^3$ Hz repetition frequency. The authors of [69] have performed detailed measurements of spatial distribution of the excited SW, and obtained a sufficiently large value of electron density in this discharge: $n_e \sim 10^{14}\,cm^{-3}$. This allowed them to conclude that this launcher for SW sustained high power microwave discharge has high efficiency.

The authors of [70] report on the elaboration of a microwave discharge, which enables the production of high-density (plasma electron density $n_e \geq 10^{11}\,cm^{-3}$), large-area (square $S \geq 5\,m^2$) flat plasma, owing to its efficient absorption of SWs. In order to supply uniform emission of microwaves over an entire plasma surface, an array of slotted waveguide antennas and a novel compact power divider have been developed. Verification of the plasma uniformity has been conducted with two-dimensional Langmuir probe measurements. Moreover, the authors have proposed general guidelines for designing slot antenna arrays for requested plasma sizes.

The authors of [71] presented results of their experiments on measurements of a dense plasma created in a microwave resonator. This is an example of specific application of a plasma discharge. An electron density up to $10^{15}\,cm^{-3}$ was obtained in a 9.4 GHz resonator cavity, which was equipped with a movable plunger at the end where the rectangular feeding waveguide was located. The second end was created by the reflecting over-dense plasma column. The microwave energy for this device was provided by a pulsed magnetron with the following parameters: 120 kW power, 1 μs pulse duration. This energy dissipated in the capillary discharge operated with 20–100 Torr helium. The discharge is sustained by propagation of microwave SWs along the protruding parts of the quartz capillarity. Analyzing the results of their experiments the authors are sure that the excitation of the SWs will provide suitable starting conditions for the detailed analysis of afterglow measurements being useful for studies of regimes of high plasma densities.

The authors of paper [72] describe a technique for the measurement of the absolute electron density in low-pressure plasmas using propagation of electromagnetic SWs along a dielectric cylinder immersed into plasma. The transmission spectrum is measured between two antennas located at the ends of the cylinder and connected to a network analyzer. The plasma transmission probe (PTP) elaborated by the authors is promising for the measurement of densities larger than $10^9\,cm^{-3}$ at relatively high gas pressure (until 1 Torr). Experimental spectra measured in RF capacitively coupled discharge in argon at various plasma densities and pressures (40 ÷ 750 m Torr) are presented. Plasma densities derived from the transmission spectra were compared with those obtained with a Langmuir probe and with a microwave 1/4-wavelength resonator at low pressure (5 ÷ 45 m Torr). The densities determined by the PTP were found to be about twice times lower compared with those obtained with other methods. The authors explained it by influence of the pre-sheath plasma density gradient, since the PTP determines the sheath edge electron density, but not the bulk value.

As it is shown in [73, 74] for simple models, surface flute waves are able to sustain gas discharges under the condition of utilization of an external magnetic field. Their eigen frequencies are in the range of the electron cyclotron resonance. Therefore they have prospects to be applied for plasma production.

## References

1. Shkvarunets, A. G., Kobayashi, S., Weaver, J., et al. (1996). Electromagnetic properties of corrugated and smooth waveguides filled with radially inhomogeneous plasma. *IEEE Transactions on Plasma Science,24*(3), 905–917.
2. Vlasov, A. N., Shkvarunets, A. G., Rodgers, J. C., et al. (2000). Overmoded GW-class surface-wave microwave oscillator. *IEEE Transactions on Plasma Science*, *28*(3), 505–560.
3. Ginzburg, N. S., Malkin, A. M., Sergeev, A. S., et al. (2001). Quasi-optical model of wave propagation near a shallow corrugated surface. *Applied Physics Letters,99*, 121505.
4. Ginzburg, N. S., Malkin, A. M., Sergeev, A. S., et al. (2012). Powerful surface-wave oscillators with two-dimensional periodic structures. *Journal of Applied Physics 100, 143510.*
5. Shin, Y. M., So, J. K., Jang, K. H., et al. (2007). Evanescent tunneling of an effective surface plasmon excited by convection electrons. *Physical Review Letters,99*, 147402.
6. Bratman, V. L., Fedotov, A. E., Makhalov, P. B., et al. (2009). Smith-Purcell frequency multiplier with synchronization of radiation from a wide electron beam. *Applied Physics Letters,94*, 061501.
7. Nusinovich, G. S., Carmel, Y., Shkvarunets, A. G., et al. (2005). The Pasotron: Progress in the theory and experiments. *IEEE Transactions on Electron Devices*, *52*(5), 845–856.
8. Bykov, V. M., Gubanov, V. P., Rostov, A. V., et al. (1991). Relativistic pulse-periodic microwave oscillators. *Proceedings of the 1st International Workshop "Strong Microwaves in Plasmas"* (Vol. 2, pp. 866–871), Russian Federation.
9. Selivanov, I. A., Fedotov, A. V., & Shkvarunets, A. G. (1991). The excitation of the fundamental mode of a coaxial plasma slow-wave structure of a PCM-amplifier. *Proceedings of the 1st International Workshop "Strong Microwaves in Plasmas"* (Vol. 2, pp. 462–467), Russian Federation.
10. Buts, V. A., Miroshnichenko, V. I., Ognivenko, V.V., et al. (1993). The excitation of a plasma-filled slow resonator by an electron beam. In N. Novgorod (Ed.), *Proceedings of the 2nd International Workshop "Strong Microwaves in Plasmas"* (Vol. 2, pp. 462–467), Russian Federation.
11. Kovalev, N. F., Kolganov, N. G., Palitsin, A. V., et al. (2000). Relativistic BWO with cyclotron selection of an asymmetric wave. In N. Novgorod (Ed.), *Proceedings of the 4th International Workshop "Strong Microwaves in Plasmas"* (Vol. 2, pp. 845–850), Russian Federation.
12. Bratman, V. L., Kalynov, Yu. K., Kolganov, N. G., et al. (1996). Cyclotron autoresonance masers and relativistic gyrotrons. In N. Novgorod (Ed.), *Proceedings of the 3rd International Workshop "Strong Microwaves in Plasmas"* (Vol. 2, pp. 745–761), Russian Federation.

13. Bratman, V. L., Denisov, G. G., Kalynov, Yu. K., et al. (2000). Novel types of cyclotron resonance devices. In N. Novgorod (Ed.), *Proceedings of the 4th International Workshop "Strong Microwaves in Plasmas"* (Vol. 2, pp. 683–702), Russian Federation.
14. Anderson, T. (2011). *Plasma antennas*. Boston: Artech House.
15. Jenn, D. C. (2003). *Plasma antennas: Survey of techniques and the current state of the art.* San Diego, CA: SPAWAR PMW 189.
16. Borg, G. G., Harris, J. H., Miljak, D. G., et al. (1999). Application of plasma columns to radiofrequency antennas. *Applied Physics Letters,74*, 3272–3274.
17. Moisan, M., Shivarova, A., & Trivelpiece, A. W. (1982). Experimental investigations of the propagation of surface waves along a plasma column. *Plasma Physics*, *20*(11), 1331–1400.
18. Borg, G. G., Harris, J. H., Martin, N. M., et al. (2000). Plasmas as antennas: Theory, experiment and applications. *Physics of Plasmas*, *7*(5), 2198–2202.
19. Alexeff, I., Anderson, T., Parameswaran, C., et al. (2006). Experimental and theoretical results with plasma antennas. *IEEE Transaction on Plasma Science*, *34*(2), 166–172.
20. Alexeff, I., Anderson, T., Farshi, E., et al. (2008). Recent results for plasma antennas. *Physics of Plasmas,15*, 057104.
21. Kumar, R., & Bora, D. (2010). A reconfigurable plasma antenna. *Journal of Applied Physics,107*, 053303.
22. Fathy, A. E., Rosen, A., Owen, H. S., et al. (2003). Silicon-based reconfigurable antennas—concepts, analysis, implementation, and feasibility. *IEEE Transactions on Microwave Theory and Techniques*, *51*(6), 1650–1661.
23. Rayner, J. P., Whichello, A. P., & Cheetham, A. D. (2004). Physical characteristics of plasma antennas. *IEEE Transactions on Plasma Science*, *32*(1), 269–281.
24. Wu, M., Wen, B. Y., & Zhou, H. (2009). Ionospheric clutter suppression in HF surface wave radar. *Journal of Electromagnetic Waves and Applications*, *23*(10), 1265–1272.
25. Russo, P., Cerri, G., & Vecchioni, E. (2011). Self-consistent analysis of cylindrical plasma antennas. *IEEE Transactions on Antennas and Propagation*, *59*(5), 1503–1511.
26. Xiong, J., Li, H., Wang, B. Z., et al. (2011). Theoretical investigation of rectangular patch antenna miniaturization based on the DPS-ENG bi-layer super-slow TM wave. *Progress In Electromagnetics Research,118*, 379–396.
27. Ye, H. Q., Gao, M., & Tang, C. J. (2011). Radiation theory of the plasma antenna. *IEEE Transactions on Antennas and Propagation*, *59*(5), 1497–1502.
28. Dynin, E. E., Kostrov, A. V., Smisnov, A. I., et al. (2000). Influence of the magnetized plasma shell on efficiency of radiation of a short antenna. In N. Novgorod (Ed.), *Proceedings of the 4th International Workshop "Strong Microwaves in Plasmas"* (Vol. 2, pp. 481–484), Russian Federation.
29. Weynants, R. R. (2009). ICRF review: From ERASMUS to ITER. *Proceedings of AIP Conference on Radio Frequency Power in Plasmas*, *1187*, 3–12.
30. Messiaen, A. M., et al. (1984). Heating in toroidal plasmas. *Proceedings of 4th Varenna-Grenoble Symposium* (Vol. 1, p. 315).
31. Ballico, R. C., & Cross, M. J. (1990). Probe measurements of ICRF Alfven surface waves in the TORTUS tokamak. *Fusion Engineering and Design*, *12*(1, 2), 197–201.
32. Murphy, A. B. (1990). Waves in the edge plasma during ion cyclotron resonance heating. *Fusion Engineering and Design*, *12*(1,2), 79–92.
33. Porkolab, M. (1990). Parametric instabilities in the tokamak edge plasma in the ion cyclotron heating regimes. *Fusion Engineering and Design*, *12*, (1,2), 93–103.
34. Noterdaeme, J.-M., ICRH-team, ASDEX-team, NI-team (1990). Experimental results on edge effects during ICRF heating of ASDEX plasmas. *Fusion Engineering and Design*, *12*(1,2), 127–201.
35. Lin, Y., Rice, J. E., Wukitch, S. J., et al. (2009). ICRF mode conversion flow drive on Alcator CMod and projections to other tokamaks. *Proceedings of AIP Conference on Radio Frequency Power in Plasmas* (Vol. 1187, pp. 57–64).
36. Lyssoivan, A., Douai, D., Koch, R., et al. (2012). Simulation of ITER full-field ICWC scenario in JET: RF physics aspects. *Plasma Physics and Controlled Fusion,54*, 074014.

37. Aliev, Yu. M., Schluter, H., & Shivarova, A. (2000). *Guided-wave-produced plasmas*. New-York: Springer.
38. Czarnecka, A., Durodie, F., Figueiredo, A. C. A., et al. (2012). Impurity production from the ion cyclotron resonance heating antennas in JET. *Plasma Physics and Controlled Fusion,54*, 074013.
39. Katanuma, I., Yagi, K., Haraguchi, Y., et al. (2011). The stability of flute modes in the GAMMA 10 A-divertor. *Plasma and Fusion Research,6*, 2403080.
40. Chen, F. F. (1974). *Introduction to plasma physics*. New York: Plenum Press.
41. Katanuma, I., Masaki, S., Sato, S., et al. (2011). Flute instability in the tandem mirror with the divertor/dipole regions. *Physics of Plasmas,18*, 112304.
42. Anders, A. (2007). Metal plasmas for the fabrication of nanostructures. *Journal of Physics D: Applied Physics,40*, 2272–2284.
43. Maier, S. A. (2007). *Plasmonics: Fundamentals and applications*. New York: Springer.
44. Beck, F. J., Polman, A., & Catchpole, K. R. (2009). Tunable light trapping for solar cells using localized surface plasmons. *Journal of Applied Physics,105*, 114310.
45. Catchpole, K. R., & Polman, A. (2009). Photovoltaic cell with surface plasmon resonance generating nanostructures, US Patent Application 20110197959. http://WWW.patentstorm.us/applications/20110197959/claims.html.
46. Green, R. J., Frazier, R. A., & Shakesheff, K. M. (2000). Surface plasmon resonance analysis of dynamic biological interactions with biomaterials. *Biomaterials,21*, 1823–1835.
47. Morrow, R., McKenzie, D. R., & Bilek, M. M. M. (2007). Electric field effects on adsorption/desorption of proteins and colloidal particles on a gold film observed using surface plasmon resonance. *Physica B,394*, 203–207.
48. Feltis, B. N., Sexton, B. A., & Glenn, F. L. (2008). A hand-held surface plasmon resonance biosensor for the detection of ricin and other biological agents. *Biosensors & Bioelectronics,23*, 1131–1136.
49. Ederra, I., Iriarte, J. C., Gonzalo, R., et al. (2011). Surface waves of finite size electromagnetic band gap woodpile structures. *Progress in Electromagnetics Research B,28*, 19–34.
50. Joannopoulos, J. D., Meade, R. D., & Finn, J. N. (1995). *Photonic crystals: Molding the flow of light*. Princeton: Princeton University Press.
51. Kawashima, S., Ishizaki, K., & Noda, S. (2010). Light propagation in three-dimensional photonic crystals. *Optics Express, 18*(1), 386–392.
52. Sozuer, H. S., & Dowling, J. P. (1994). Photonic band calculations for woodpile structure. *Journal of Modern Optics,43*, 231–234.
53. Gonzalo, R., Ederra, I., Mann, C. M., et al. (2001). Radiation properties of terahertz dipole antenna mounted on photonic crystal. *Electronics Letters, 37*(10), 613–614.
54. Ederra, I., Gonzalo, R., & Alderman, B. E. J. (2008). Electromagnetic band-gap based planar imaging array for 500 GHz. *IEEE Transactions Microwaves Theory Technology, 56*(11), 2256–2265.
55. Weily, A. R., Esselle, K. P., & Sanders, B. C. (2003). Photonic crystal horn and array antennas. *Physical Review E, 68*(1), 016609.
56. Saxler, J., Rivas, J. G., & Janke, C. (2004). Time-domain measurements of surface plasmon polaritons in the terahertz frequency range. *Physical Review B,69*, 155427.
57. Gerasimov, V. S., Cherkassky, V. S., & Knyazev, B. A., et al. (2012). Study of surface plasmons travelling along straight and curved metal-dielectric interfaces: Experiment and theory. *37th International Conference on Infrared, Millimeter and Terahertz Waves*, Wollongong, Australia, September 23–28, Tue-A-4-4.
58. Ilin, N. V., Kondratiev, I. G., & Smirnov, A. I. (2009). Planar wave guiding systems based on resonant meta materials. In N. Novgorod (Ed.), *Proceedings of the 7th International Workshop "Strong Microwaves in Plasmas"* (Vol. 2, pp. 550–554), Russian Federation.
59. Moisan, M., & Zakrzewski, Z. (1991). Plasma sources based on the propagation of electromagnetic surface waves. *Journal of Physics D: Applied Physics,24*, 1025–1048.
60. Margot-Chaker, J., Moisan, M., Chaker, M. et al. (1989). Tube diameter and wave frequency limitations when using the electro-magnetic surface wave in the m=1 (dipolar) mode to sustain a plasma column. *Applied Physics, 66*(9), 4134–4148.

61. Sugai, H., Ghanashev, I., & Nagatsu, M. (1998). High-density flat plasma production based on surface waves. *Plasma Sources Science and Technology,7*, 192–205.
62. Margot-Chaker, J., Moisan, M., & Teichmann, J. (1991). A new approach to the development of ECR plasma sources. *Proceedings of the 1st International Workshop "Strong Microwaves in Plasmas"* (Vol. 1, pp. 473–478), Suzdal, Russian Federation.
63. Shibkov, V. M., Chernikov, V. A., Ershov, A. P., et al. (2003). Surface microwave discharge in a supersonic airflow. In N. Novgorod (Ed.), *Proceedings of the 5th International Workshop "Strong Microwaves in Plasmas"* (Vol. 2, pp. 737–742), Russian Federation.
64. Shibkov V. M., & Shibkova L. V. (2011). Surface microwave discharge in air. In N. Novgorod (Ed.), *Proceedings of the 8th International Workshop "Strong Microwaves and Terahertz Waves"* (pp. 341–342) Sources and Applications, Russian Federation.
65. Kossyi, I. A. (2011). *New microwave plasma sources based on the latest advances of fundamental gas—discharge physics.* Ibid, 25.
66. Ivanov, O. A., & Koldanov, V. A. (2000). Self-consistent model of the pulsed SW-discharge in hydrogen. In N. Novgorod (Ed.), *Proceedings of the 4th International Workshop "Strong Microwaves in Plasmas"* (Vol. 1, pp. 319–324), Russian Federation.
67. Skovoroda, A. A., & Zhyl'tsov, V. A. (1996). The microwave discharge in a plasma neutralizer of ITER injector. In N. Novgorod (Ed.), *Proceedings of the 3rd International Workshop "Strong Microwaves in Plasmas"* (Vol. 1, pp. 368–373), Russian Federation.
68. Skovoroda, A. A., Kulygin, M. V., Smirnov, V. P., et al. (2006). Application of steady state microwave discharges in nuclear technology. In N. Novgorod (Ed.), *Proceedings of the 6th International Workshop "Strong Microwaves in Plasmas"* (Vol. 2, pp. 647–656), Russian Federation.
69. Vikharev, A. L., Gorbachev, A. M., Ivanov, O. A., et al. (1996). Long plasma columns with high electron densities produced by pulsed surface waves discharges. In N. Novgorod (Ed.), *Proceedings of the 3rd International Workshop "Strong Microwaves in Plasmas"* (Vol. 1, pp. 395–400), Russian Federation.
70. Ishijima, T., Toyoda, H., Takanishi, Y., et al. (2011). Design of large-area surface wave plasma excited by slotted waveguide antennas with novel power divider. *Japanese Journal of Applied Physics,50*, 036002.
71. Boas, C., Himmel, G., & Hirsch, S. (1993). Diagnostics on the plasma created in a pulsed microwave resonator. In N. Novgorod (Ed.), *Proceedings of the 2nd International Workshop "Strong Microwaves in Plasmas"* (Vol. 1, pp. 304–308), Russian Federation.
72. Dine, S., Booth, J.-P., Curley, G. A., et al. (2005). A novel technique for plasma density measurement using surface-wave transmission spectra. *Plasma Sources Science and Technology,14*, 777–786.
73. Girka, V., Girka, I., & Pavlenko, I. (2001). Electrodynamic model of the gas discharge sustained by azimuthal surface waves. *Contributions to Plasma Physics. 41*(4), 393–406.
74. Girka, V., Girka, I., Girka, A., et al. (2011). Theory of azimuthal surface waves propagating in non-uniform waveguides. *Journal of Plasma Physics, 77*(4), 493–519.

# Index




V. Girka et al., *Surface Flute Waves in Plasmas*, Springer Series on Atomic, Optical, and Plasma Physics 79, DOI: 10.1007/978-3-319-02027-3,

FSC
www.fsc.org
MIX
Papier | Fördert gute Waldnutzung
FSC® C083411

Zeitfracht Medien GmbH
Ferdinand-Jühlke-Straße 7
99095 Erfurt, Deutschland
produktsicherheit@kolibri360.de